Human-Centered Design for Mining Equipment and New Technology

Human-Centered Design for Mining Equipment and New Technology

By
Tim Horberry, Robin Burgess-Limerick, and
Lisa Steiner

CRC Press
Taylor & Francis Group
Boca Raton London New York

CRC Press is an imprint of the
Taylor & Francis Group, an **informa** business

MATLAB® is a trademark of The MathWorks, Inc. and is used with permission. The MathWorks does not warrant the accuracy of the text or exercises in this book. This book's use or discussion of MATLAB® software or related products does not constitute endorsement or sponsorship by The MathWorks of a particular pedagogical approach or particular use of the MATLAB® software.

CRC Press
Taylor & Francis Group
6000 Broken Sound Parkway NW, Suite 300
Boca Raton, FL 33487-2742

© 2018 by Taylor & Francis Group, LLC
CRC Press is an imprint of Taylor & Francis Group, an Informa business

No claim to original U.S. Government works
Printed on acid-free paper

International Standard Book Number-13: 978-1-138-09514-4 (Hardback)
International Standard Book Number-13: 978-1-138-09520-5 (Paperback)

Library of Congress Cataloging-in-Publication Data

Names: Horberry, Tim, author. | Burgess-Limerick, Robin, author. | Steiner, Lisa J., author.
Title: Human-centered design for mining equipment and new technology / Tim Horberry, Robin Burgess-Limerick, Lisa Steiner.
Description: Boca Raton : Taylor & Francis, a CRC title, part of the Taylor & Francis imprint, a member of the Taylor & Francis Group, the academic division of T&F Informa, plc, [2018] | Includes bibliographical references and index.
Identifiers: LCCN 2017048091 | ISBN 9781138095144 (hardback : acid-free paper) | ISBN 9781138095205 (pbk. : acid-free paper) | ISBN 9781315105772 (ebook)
Subjects: LCSH: Mine safety. | Human engineering. | Mining machinery--Design and construction. | Human-machine systems.
Classification: LCC TN295 .H739 2018 | DDC 622.028/4--dc23
LC record available at https://lccn.loc.gov/2017048091

Visit the Taylor & Francis Web site at
www.taylorandfrancis.com

and the CRC Press Web site at
www.crcpress.com

Contents

List of Tables

List of Figures

Foreword

I am very pleased to have been asked to write this foreword for *Human-Centered Design for Mining Equipment and New Technology*. This book will most certainly help the reader to understand the importance of focusing on "the human and the tasks they do" when designing, developing, and upgrading mining equipment and the value in using a wide cross-section of inputs, including from the end users, while going through these processes. My personal passion is improving environment, health and safety in workplaces—HCD clearly has an important role to play there, but the opportunities to significantly improve efficiency and productivity are also evident.

When supplying to a truly global marketplace, Sandvik and other original equipment manufacturers (OEMs) are faced with an enormous challenge just to meet the large number of different and regularly changing regulations and standards. Meeting these existing requirements is seen as a "minimum requirement" by almost all organizations that purchase mining equipment. However, many of these organizations, often supported or pressured by other external stakeholders, now require OEMs to go "beyond compliance" and look for innovative ways to produce equipment that is safer, more environmentally friendly, and more productive. HCD is an approach that has been used to good effect in other industries, and I hope this book encourages OEMs and the mining industry to use HCD more broadly.

The mining industry is in a very exciting period of change, and OEMs will need to be both "forward looking" and innovative to prosper. I believe that using HCD as part of this change will help move the mining industry to its often-stated vision of Zero Harm.

Stuart Evans
Head of EHS, Sandvik Group
Vice President EHS, Sandvik Mining & Rock Technology

Acknowledgements

The authors are grateful to colleagues at Monash University, University of Queensland, NIOSH PMRD, and elsewhere for their assistance. The findings and conclusions in this book are those of the authors and do not necessarily represent the views of the National Institute for Occupational Safety and Health (NIOSH), Centers for Disease Control, USA. Mention of any company or product does not constitute endorsement by NIOSH.

We specifically acknowledge Jessica Merill who was on a student program at NIOSH PMRD as the lead writer of the "dust hopper" case study in Chapter 5. We also acknowledge Danellie Lynas from the University of Queensland as a co-author of Section 4.3.4.6. The content for the book was largely from two research projects: *'Establishing a Human Centered Design Approach for Mining Technologies'* contract number 200-2014-M-59067, and *'Human Centered Design Case Studies,'* contract number 200-2015-M-6294 with the Centers for Disease Control and Prevention, USA. In addition, the proximity detection case study arose from Australian Coal Association Research Program (project C24028). We thank the AusIMM for giving permission to reproduce a conference paper about underground loader automation at CMOC Northparkes we previously published in the 13th AusIMM Underground Operators Conference 2017. We also acknowledge other relevant research funding received by the authors from NIOSH PMRD, the European Commission (Marie Curie Fellowships), EMESRT, CRC Mining, Australian Coal Association Research Program, and Coal Services Health and Safety Trust. Finally, we extend our greatest thanks to our families for their continuing support.

Authors

Professor Tim Horberry leads the human factors team at Monash University Accident Research Centre in Australia. He recently was a senior visiting researcher on a "safety in design ergonomics" Fellowship at Cambridge University in England. Tim's background is in human factors, safe design, and transport/mining safety.

Professor Robin Burgess-Limerick is a professorial research fellow at the Minerals Industry Safety and Health Centre, Sustainable Minerals Institute, The University of Queensland, Australia. Robin has been a qualified ergonomist for over 20 years, and is a past-president and fellow of the Human Factors and Ergonomics Society of Australia.

Dr Lisa Steiner is associate director of science at the Pittsburgh Mining Research Division at the National Institute for Occupational Safety & Health, Pittsburgh, USA. Lisa has a background in industrial engineering and has over 20 years of experience in mining and ergonomics. She completed her PhD, "Reducing Underground Coal Roof Bolting Injury Risks through Equipment Design," at the University of Queensland, Australia, in 2014.

Glossary of Terms and Acronyms

EMESRT: Earth Moving Equipment Safety Round Table

The Earth Moving Equipment Safety Round Table (EMESRT) is a global initiative of mining companies. EMESRT engages with key mining industry manufacturers to advance the design of the equipment to improve safe operability and maintainability.

EDEEP: EMESRT Design Evaluation for EME Procurement

EMESRT developed a task-based design evaluation process known as the EMESRT Design Evaluation for EME Procurement—EDEEP (Burgess-Limerick, Joy, Cooke, and Horberry, 2012).

FDA: US Food and Drug Administration

HCD: human-centered design

An approach to systems design and development that aims to make systems more usable by focusing on the use of the system and applying human factors/ergonomics and usability knowledge and techniques (ISO 9241:210, 2010).

HF: human factors/ergonomics

The scientific discipline concerned with the understanding of the interactions among people and the other elements of a work system, and the profession that applies theory, principles, data, and methods to design in order to optimize human well-being, safety, and overall system performance (Horberry, Burgess-Limerick, and Steiner, 2011).

ISO: International Organization for Standardization

MINER Act: The US Mine Improvement and New Emergency Response Act of 2006 (MINER Act, 2006).

MSHA: Mine Safety and Health Administration (US Department of Labor)

OEM: original equipment manufacturer

PMRD: Pittsburgh Mining Research Division (Health and Human Services, Center for Disease Control)

PtD: prevention through design

PtD involves all of the efforts to anticipate and design out hazards to workers in facilities, work methods and operations, processes, equipment, tools, products, materials, new technologies and the organization of work (Schulte et al., 2008).

Safe design: See PtD. Also sometimes referred to as 'safety in design' or 'prevention through design', this approach emphasises safety by original design to remove hazards and reduce risks. It reduces the need for safety by procedure, other administrative controls, or trial and error (Horberry and Burgess-Limerick, 2015).

MATLAB® is a registered trademark of The MathWorks, Inc. For product information, please contact:

The MathWorks, Inc.
3 Apple Hill Drive
Natick, MA 01760-2098 USA
Tel: 508-647-7000
Fax: 508-647-7001
E-mail: info@mathworks.com
Web: www.mathworks.com

1 Why HCD for Mining Equipment?

1.1 OVERVIEW

This book focuses on human-centered design (HCD) and outlines the benefits of this approach for mining equipment and new technology. HCD is widely used in the design of medical equipment, road vehicles, and consumer products. It is a process that aims to make equipment and systems more usable and acceptable by explicitly focusing on the end users, their tasks, their work environment, and the use context. A key requirement of human-centered design is for users and other stakeholders to be involved throughout the design and development of the equipment or system. However, to date, HCD has not been widely applied to the design, development, and deployment of equipment or new technology in the mining industry.

Human-centered design links closely with the aims of the NIOSH Prevention through Design initiative for "designing out" hazards. In the USA, HCD is supported by the MINER Act (2006) for the development of new human-centered mining technologies and the US Board on Human Systems Integration (2013) for human-centered self-escape systems. The global mining industry is currently seeing a rapid growth in the design and deployment of smart devices, automation, tele-operation systems and new mining equipment: it is particularly important that such systems are designed to be safe, effective, and usable.

This book shows that HCD should become a key design and development process in mining: effective design for human use. It shows that human-centered design can result in fewer operator errors, decrease training costs, promote better system usability, and give improved operator acceptance of systems. In mining, HCD can help minimize design issues like operator overload from too many warnings in truck cabs, or self-rescue equipment not being deployed because of poor device usability. Human-centered design in mining is valuable, necessary, and timely.

1.2 WHAT IS HCD?

Human-centered design focuses design on end users, their tasks, and the environmental context in which tasks are performed. HCD requires the involvement of operators, maintainers, and other stakeholders throughout the design, development, and implementation of equipment or technology: by identifying human-related design requirements and actively involving end users throughout the entire design process. Of course, human-centered design does not operate in a vacuum: it balances the needs of users with the financial, technical, and organizational constraints that are always present (Sharples et al., 2016). As will be seen in examples throughout this book, employing HCD improves usability, productivity, and safety for those who

1

operate, maintain, and work near mining equipment. However, despite the examples of success, human-centered design techniques and processes are not universally applied to the design or deployment of mining equipment or new technology.

The most relevant ISO standard in this area (ISO 9241 Part 210: Human-centered design for interactive systems, 2010) is now becoming well accepted in many domains. A successor to ISO 13407:1999, it can also easily be applied to mining equipment and new technologies. Therefore, the ISO 9241-210 HCD definition is used as the basis for human-centered design throughout this book:

> Approach to systems design and development that aims to make interactive systems more usable by focusing on the use of the system and applying human factors/ergonomics and usability knowledge and techniques. (ISO 9241-210, 2010)

1.3   APPLICATION OF HCD TO EQUIPMENT AND NEW TECHNOLOGIES USED IN THE MINERALS INDUSTRY

The preceding ISO definition of human-centered design is applicable to mining/minerals industry equipment and new technologies. For example, with respect to a new technology such as a proximity detection system for a large mining vehicle, the ISO 9241-210 description could be operationally defined as:

> An approach to proximity detection system design and development for mobile mining equipment that aims to make the system safer and more usable by focusing on the actual mine site use of the proximity detection system and applying human factors/ergonomics and usability knowledge and techniques.

ISO 9241-210 does not have an explicit safety focus: instead, it is mainly concerned with usability and effectiveness. However, because safety is a key driver in the minerals industry, the working definition of mining human-centered design presented here explicitly also considers safety as well as usability and performance issues.

1.4   KEY POINTS OF HCD

As Rouse (2007) notes, HCD has three key objectives, especially in the earlier stages of design. These are: to **enhance human abilities** (e.g., in hazard recognition), to **help overcome human limitations** (e.g., the inclination to make errors) and to **foster human acceptance** (e.g., of new mining technologies).

It should be noted that human-centered design is wider and more embracing than the term "user-centered design" because it considers the impact on other stakeholders who might not typically be considered as users, such as maintenance staff.

Similarly, HCD is not exactly the same as human factors. As the ISO 9241-210 definition states, HCD applies human factors knowledge and techniques as part of the design process to make systems and products usable and useful. The following ten key points characterize HCD:

1. It is often now used as an umbrella term, covering other terms such as "user-centered design," "interaction design," and "design for product experience."

2. Similarly, it uses knowledge and techniques drawn from human factors, ergonomics, and usability engineering, but is not the same as these topics.
3. It is different from other design movements such as technology-driven design or sustainable design. Technical innovations or environmental impacts are not the key focus of human-centered design.
4. The focus is upon making systems and products more usable, useful, and acceptable and less likely to result in adverse safety or health effects.
5. The aim is to bring benefits such as improved productivity, user well-being, accessibility, fewer errors, and reduced harm.
6. It fits the product, system, or interface to the user, not vice versa. It tries to optimize the system/equipment around the ways users can, want, or need to use it, rather than forcing them to change their behavior to accommodate the product.
7. It requires an understanding of the users, their tasks, and the environment/use context.
8. Users and other stakeholders should be involved throughout design and development.
9. The design is iterative and driven by human-centered evaluation criteria (e.g., usability/safety).
10. The needs, wants, and limitations of all people who may interact with a product or equipment are given explicit attention at each stage of the design process.

1.5   THE BENEFITS OF HCD . . . AND THE FAILURES OF NOT TAKING A HUMAN-CENTERED APPROACH

There are considerable benefits to be obtained through using HCD in mining. The examples presented later in this book show that successful products can be developed in the minerals industry and in related domains by using human-centered design approaches. Work by Horberry et al. (2011) characterized the benefits of using a human-centered approach to mining equipment as being either safety- and health-related, or productivity- and efficiency-focused. Similarly, Kujala (2003) in a review of cost/benefit evidence identified the following benefits that could be obtained when there is extensive user involvement during design:

- Increased user productivity/fewer errors
- Increased sales
- Decreased training costs
- Decreased user support
- More accurate end-user requirements and better system usability
- Fewer costly system features that are unwanted or irrelevant to the task
- Improved operator acceptance and understanding of the system/product

Of course, not all the mining equipment or new technologies developed using HCD approaches are guaranteed to have all these benefits. Nonetheless, the available evidence about human-centered design from mining and elsewhere indicates that it has a positive effect overall (Kujala, 2003; Horberry et al., 2011).

1.5.1 Participatory Ergonomics and Human-Centered Design

One effective way of encouraging human-centered design (and particularly, redesign) of work is the implementation of participatory ergonomics programs. Participatory ergonomics means actively involving workers in developing and implementing workplace changes that will improve productivity and reduce risks to safety and health—or, as Wilson (1995) put it, the "involvement of people in planning and controlling a significant amount of their own work activities, with sufficient knowledge and power to influence both processes and outcomes to achieve desirable goals." The underpinning assumption is that workers are the experts, and that given appropriate knowledge, skills, tools, facilitation, resources, and encouragement, they are best placed to identify and analyze problems and to develop and implement solutions, which will be both effective in reducing injury risks and improving productivity, and will be acceptable to those impacted (Brown, 2005).

Participatory ergonomics programs have been implemented across a large range of industries and organizations, including mining (Burgess-Limerick et al., 2007; Torma-Krajewski et al., 2007) and many others. Perhaps as a consequence of the diverse settings in which programs have been implemented, and the need for programs to "fit" each organization or situation, there are many variations in the program characteristics, such as the degree and nature of participation, the extent of expert facilitation and assistance provided, the nature and extent of training provided to teams (in ergonomics methods and team work), and the tools employed to assist teams in identifying issues and developing solutions.

Participatory ergonomics is reported to have a range of benefits in addition to the reduction in injury risks, such as an improved flow of useful information within an organization, an improvement in the meaningfulness of work, more rapid technological and organizational change, and enhanced performance (Brown, 1993; Haims and Carayon, 1998; Haines and Wilson, 1998). As well as developing more effective solutions, the use of participative ergonomics techniques to derive solutions is believed to result in greater "ownership" by those affected, leading to a greater commitment to the changes being implemented.

Although some research has demonstrated the significant effects of implementing a participatory ergonomics program on physical risk factors (e.g., Straker et al., 2004) or productivity (e.g., Vink et al., 1995), most evaluations have focused on direct health effects. The outcomes of individual evaluations are mixed. Silverstein and Clark (2004) noted this variability, concluding that participatory ergonomics programs were "often, but not always successful." Cole et al. (2005) reviewed 10 evaluations of the health effects of participatory ergonomics programs, concluding that the studies provided limited evidence that participatory ergonomics programs can have a positive impact on musculoskeletal injury symptoms and compensation costs. More encouragingly, Rivilis et al. (2008) concluded that the "12 studies that were rated as 'medium' or higher provided partial to moderate evidence that PE interventions have a positive impact on: musculoskeletal symptoms, reducing injuries and workers' compensation claims, and a reduction in lost days from work or sickness absence."

More recent evaluations not included in these reviews have also demonstrated mixed results; however, Cantley et al. (2014) reported positive outcomes from a six-year evaluation of a large-scale participatory ergonomics process at a multisite aluminum manufacturer. Control measure implementation targets were set by senior management, and the evaluation noted 204 control measures implemented across 123 jobs at 17 study sites, affecting the work of 14,540 workers. Jobs in which control measures were introduced were associated with significantly lower musculoskeletal injury risk, and the authors concluded that the study "provides evidence that a systematic approach to ergonomic hazard identification, quantification, and control implementation, in conjunction with requirements to establish an ergonomic process at each manufacturing plant, may be effective in reducing the risk of MSD and acute injury outcomes among workers in targeted jobs" Cantley et al. (2014).

The mixed nature of evaluations is perhaps unsurprising given the diversity of program designs and the variety of organizational characteristics and contexts in which program implementation has been attempted. It is reasonable to conclude that, although participatory ergonomics programs have the potential for positive health benefits, there are many potential barriers, and success has not always been achieved.

Organizations that are less hierarchical, have good labor relations, have a tradition of consultation in other areas, maintain good communication channels, and have job designs emphasizing personal control are likely to most easily adopt and benefit from a participatory ergonomics program. The commitment of management, at all levels, to the program is the most important factor contributing to the probability of success (Liker et al., 1989; Brown, 2005; Dixon et al., 2009). Senior management commitment is essential for ensuring that adequate resources are available, including the provision of time for team members to participate in training and intervention activities, and the approval of the expenditures required to implement workplace changes (Haines and Wilson, 1998). Constraint on the availability of such resources (both time and money) have been noted as providing significant barriers to success in some studies (e.g., Torma-Krajewski et al., 2007).

It is also important to ensure that middle managers within the organization maintain commitment in the face of inevitable production pressures. The challenge of achieving this is well described by Dixon et al. (2009), who investigated the implementation of three participatory ergonomics programs:

> While senior management in all sites was supportive at the outset of the process, it was middle management and supervisors who, for the most part, had to deal with the pragmatic issues around maintaining production once the intervention program was in progress. Given the pressures they faced, it is not surprising that securing their support was an ongoing challenge. Additionally, there were variations across the sites in the degree to which senior management's initial commitment was sustained throughout the course of the ergonomics program, and importantly, backed up by active intervention when the program encountered resistance (p. 67).

Programs are likely to be most successful when senior management dictates implementation targets for task redesigns across the organization (e.g., Cantley et al., 2014; Dennis et al., 2015; Pazell et al., 2016).

Ensuring genuine participation by team members is the next hurdle that some workplaces encounter, depending on the history of relationships between management and workers. As well as management being committed, workers need to believe this to be the case. The role of the facilitator of the ergonomics program may sometimes need to extend to facilitating communication between management and workers. The presence of a site champion to drive the process has been considered important, and staff turnover, which disrupts the site champions' ongoing contributions, is a threat to program sustainability, as is general instability within the workplace or industry generally.

It is important to ensure that worker participation continues throughout the development and revision of redesign activities, as well as the initial solution ideation stages. It is also important to ensure that internal experts (e.g., engineers) are included in the process when appropriate, as well as those potentially affected by proposed design changes (e.g., maintenance staff).

1.5.2 FAILURES WITH EQUIPMENT/TECHNOLOGY BY NOT CONSIDERING THE HUMAN ELEMENT

Conversely, many failures with equipment/technology have arisen in other domains by not considering the human element in design and operation. Burgess-Limerick et al. (2011) summarized situations in which human system integration/human-centered design was insufficient. Such detailed data does not exist for mining, but focusing on other comparable, high hazard work domains (military and aviation), these include:

- **The Black Hawk helicopter**. Deficiencies of human factors, manpower, personnel, and training were identified during the "reverse engineering" of the Black Hawk helicopter acquisition program.
- **Aquila remotely piloted vehicles**. Many human element problems discovered during testing and development of the US Army's Aquila remotely piloted vehicle led to the cancellation of the program.
- **Patriot air and missile defense units** were involved in two incidents occurring during Operation Iraqi Freedom (18% of engagements), which resulted in fatalities of allied forces. It was concluded by the authors that the causes of operator errors can be traced to decisions made by designers and others responsible for the development of the system over 25 years. The dominant mode of control changed from manual to supervisory control as increasing levels of automation were added. However, the operators' role change was not reflected in design and evaluation, or in training practices.
- **Catastrophic events have been caused by failure to consider human capabilities**, including the downing of Korean Air Lines Flight 007, which strayed into Soviet air space; the Three Mile Island nuclear accident; the downing of Iran Air Flight 655 by the USS Vincennes; the Bhopal release of methyl isocynate; and the 1972 crash of a Lockheed L-1011 in the Florida everglades.
- **Salyut 11**. The fatal decompression of Salyut 11 is an example of a failure to consider human capabilities in design.

- **Anzac-class frigates**. The operations room of the Anzac-class frigates required redesign to correct deficiencies that resulted in poor situation awareness for the command team, space restrictions, excessive reach distances, and visibility issues.
- **Seasprite helicopter cockpit** design issues, with detrimental operational consequences, cost an estimated A$100–200m to rectify.
- **F-111**. An insufficient focus on "the incorporation of OHS concerns into engineering design" was also identified as a factor which contributed to the chemical exposure of Australian Air Force maintenance workers during F-111 fuel tank maintenance, leading to recommendations by the Board of Inquiry that "occupational health and safety should be integrated into the engineering change management process. This means, in particular, that designs should undergo a risk management process" and that "the Air Force should review its acquisition policies to ensure that suppliers have systematically identified the hazards posed to personnel who use or maintain the equipment and, as far as possible, designed out these hazards".

Together, these brief examples show that failure to properly consider the human element in equipment/system design, development, and deployment can lead to unsafe and expensive outcomes.

1.6 HCD USE SCENARIOS IN MINING AND THE MINERALS INDUSTRY

This book is intended to outline the general benefits of a mining HCD approach for original equipment manufacturers (OEMs), technology developers, regulators, researchers such as those at the Pittsburgh Mining Research Division at NIOSH in Pittsburgh, PA, and, of course, mine sites. As explored throughout the book, different job roles have different requirements for human-centered design. For example, a mine health and safety manager would be particularly focused on the health and safety of their workforce when using mining equipment, whereas an equipment designer is also concerned with developing an effective, acceptable, and successful product.

One distinction that is sometimes made is between original design and redesign: in other words, developing something new as opposed to modifying it and the work system around it to improve safety and efficiency (Horberry and Burgess-Limerick, 2015). However, it is not always clear-cut where original design finishes and redesign takes over. Using mining haul trucks as the example, the design of a new version of a truck might be "original" but still an iteration of a previous design of a haul truck. Further, the base level of a new haul truck is often modified by the OEM, dealers, and the mine site before it is put into service: such as fitting additional access/egress systems. Even when the truck is in service it often is subject to modification, for example, the fitting of new "smart" technologies such as proximity detection systems. In this book, the full equipment "life cycle" will be considered to include design, redesign, modification and procurement from a human-centered design perspective.

1.7 HCD EXAMPLE: THE NEED FOR HCD TO HELP DEVELOP SELF-ESCAPE TECHNOLOGY

The need for HCD in the development of self-escape technology provides a tangible example of the material to be discussed. The Mine Improvement and New Emergency Response Act of 2006 (MINER Act) promulgated in the USA noted the need for technologies to be developed that could enhance self-escape from underground mines. In 2013, the US National Academies Press publication *Improving Self-Escape for Underground Coal Mines* recommended that technologies to enhance self-escape should be developed using human-centered design principles. It was noted that although technology advances might be "awe inspiring," their integration into mining operations might not always be in the best interests of the miner. For example:

- A miner has only a certain amount of strength and space to carry such self-escape devices. If the devices are overly heavy or bulky, then miners might not accept them or be able to escape.
- In emergencies, miners should not be required to monitor and understand data from multiple devices and multiple sources (due to information-overload concerns). Also, if devices are not easy to understand then they may be trusted less and perhaps not used.

The US National Academies Press report recommended that new self-escape technologies focus on key functions including tracking miner position, communication, atmosphere monitoring, and vision enhancement. It recommended designing devices that combine different functions to help reduce physical exertion from carrying several different single-function devices underground, and to help reduce excessive attentional demands from multiple devices for the wearer.

Using human-centered design principles could help technology to meet these objectives by specifically considering the miners, their tasks, and the mine environment where the technology would be used. The usability of such devices is particularly important as the Royal Commission's Report on the Pike River mining disaster in New Zealand (Pike River, 2012) found that some self-rescuer equipment did not seem to be working properly—so operators wrongly discarded it. Likewise, potentially faulty self-rescuers were implicated in the Sago Mine (West Virginia, US) disaster (Kowalski-Trakofler and Vaught, 2012).

How HCD might actually be used to help develop effective self-rescue technology is explored in more detail throughout this book. This includes how different human-centered design processes can be used at different stages of the self-rescue technology life cycle: from original concept through to design, deployment, modification, and disposal.

1.8 STRUCTURE OF THE BOOK

The purpose of this book is not to teach designers how to design mining equipment, nor is it to guide mine sites in running their operations in a safe and efficient manner. Rather, the key purpose is to assist mining equipment and new technology to be designed and deployed from a human-centered perspective to assist with safety and efficiency. To achieve this, the book is divided into six chapters:

- In this first chapter, HCD is overviewed, benefits are described, and the key aspects of the approach are introduced.
- The second chapter goes into more depth to outline the key principles, processes, and tools for human-centered design: these are then summarized, and an example of designing adequate underground lighting is given to help reinforce the points made.
- In the third chapter, the current status of mining HCD is examined. This includes analyzing how much human-centered design has previously been undertaken in the design of mining equipment, describing the key centers of excellence in mining HCD, and outlining some of the barriers for human-centered design. Examples of previously successful HCD initiatives are then introduced and a description is given about how human-centered design can be employed to develop fit-for-purpose self-escape technology.
- Chapter 4 provides human-centered design educational material. It describes the mining equipment life cycle and explains how to initiate mining HCD. It then presents an educational guide for undertaking mining human-centered design at different points in the equipment life cycle (for example, during design or after the equipment has been deployed in the mine site).
- Chapter 5 presents detailed case studies of human-centered design. Building on the educational material, it gives comprehensive details of HCD in different facets of mining and the minerals industry. These include "traditional" mining equipment such as roof bolters as well as cutting-edge underground automation applications.
- Finally, Chapter 6 draws together the content, analyzes the lessons learnt from the case studies and other examples, and examines what works in mining HCD. Then, the future use of human-centered design in automated mining is presented, the use of human-centered design within wider human system integration is outlined. Overall conclusions are given: what is needed for the minerals industry to better embrace HCD.

We believe that this book is timely because no available publication fully describes mining HCD, provides multiple case studies, and gives educational material. There is a growing realization within the mining community and related domains of the importance of human element issues: this includes during the design and deployment of new mining technology and automation: from assistance systems through to full automation of major mining functions. The intention of this "how to" book is to make human-centered design attractive, accessible and practical for mining equipment manufacturers, technology developers, mine site personnel, applied human factors researchers, safety scientists, and regulators.

2 Principles, Processes, and Tools for HCD

2.1 OVERVIEW

This chapter outlines three related areas of human-centered design: key principles, examples of design processes, and the tools available for HCD data collection and evaluation. For these three areas, the material is drawn from mining equipment/new technologies with HCD where possible. However, generic human-centered design principles, processes, and tools are also reviewed where they can offer guidance for mining. The HCD processes noted in the most relevant ISO documents for mining human-centered design are considered in detail. Thereafter, a summary of the essential principles, processes, and tools for mining human-centered design is given.

We conclude that, if legislative requirements around the world were to oblige it, the use of a human-factors summary report by mining technology/equipment manufacturers to demonstrate how they have addressed key HCD issues (including principles, processes, and tools) could be beneficial. This approach is similar to that employed by the US FDA for human-centered medical device designs and some mining equipment in Australia (e.g., as required in MDG-15 (2002) where a supplier should provide to the owner a statement of compliance that includes a risk assessment).

2.2 KEY PRINCIPLES OF HCD

The most relevant standard in the area, ISO 9241-210 (2010), proposes that a human-centered approach to design has substantial economic and social benefits for users, employers, and suppliers. The standard notes that highly usable systems, equipment, and products tend to be more successful both technically and commercially. Support and training costs are reduced when users can understand and use products without additional assistance. ISO 9241-210 (2010) notes that in many countries, both employers and suppliers have legal obligations to protect users from risks to their health and safety, so employing human-centered methods can help reduce these risks.

ISO 9241-210 gives a set of principles for HCD: the design is based upon an explicit understanding of users, tasks, and environments; users are involved throughout design and development; the design is driven and refined by user-centered evaluation; the process is iterative; the design addresses the whole user experience; and the design team includes multidisciplinary skills and perspectives.

Other researchers, such as Gulliksen et al. (2001, 2003), Horberry et al. (2011), and Travis (2009), in mining and elsewhere have given similar lists of key HCD principles. To synthesize these sources results in five key human-centered design principles that are relevant to the minerals industry:

1. It fits the product, system, or interface to people, not vice versa. It tries to optimize the system/equipment around the ways users/maintainers can, want, or need to interact, rather than forcing them to change their behavior to accommodate the product, system, or equipment.
2. It requires an understanding of the people who will interact with the product system or equipment, their tasks, and the environment/use context.
3. Users/maintainers and other stakeholders should be involved throughout design and development.
4. The design is iterative and driven by human-centered evaluation criteria (e.g., usability).
5. The needs, wants, and limitations of people who will interact with the product, system, or equipment are given explicit attention at each stage of the design process.

2.3 HCD PROCESSES

As previously noted by Horberry et al. (2011), on the broadest level, one of the main things all equipment and technology designers have in common is a design process. That is, the activities while conceptualizing, designing, and building a new piece of equipment. As will be seen in Chapter 3, though, the amount of HCD employed during the equipment design processes varies.

2.3.1 AN ITERATIVE PROCESS

A key thing to note about the human-centered design process is that it is iterative—it is not a fully linear process in which a single idea remains unchanged from concept through to production (ISO 9241-210, 2010). Indeed, the complexity of new technology interaction in mining means that it is impossible to specify completely and accurately every detail of every aspect of the interaction at the beginning of development. Many of the needs and expectations of end users/maintainers and other stakeholders that will impact on the design of the interaction only emerge in the course of development and deployment, as the designers refine their understanding of the users/maintainers and their tasks, and as mine site end users are better able to express their needs in response to potential solutions (ISO 9241-210, 2010). So the process, especially for complex mining equipment, usually involves revisiting, refining, and changing design ideas, and then testing, evaluating, and refining (Horberry, 2012).

Because many equipment designers and mine site engineers still view human constraints to be less significant than the technical challenges of, say, equipment reliability, there is a tendency to not systematically consider human factors early in the equipment life-cycle process, and it is common to see HCD concerns being passed from one phase to the next (Horberry et al., 2011). For example, if interface design requirements are not captured, then subsequent inadequacies cannot be resolved in later design phases. This is a particular problem for mining equipment that requires considerable human intervention in its maintenance and operation (e.g., for low-seam underground equipment).

Relevant HCD processes are reviewed in the following material. This begins by first reviewing the general human-centered design framework specified in ISO 9241-210 (2010) that is relevant to mining HCD.

2.3.2 KEY ISO FRAMEWORK TO SUPPORT MINING HCD

ISO 9241-210 (2010) provides a general framework and process for human-centered design activities that are applicable to mining. Applying this to mining equipment and to new technology, after the need for developing a mining system or product has been identified and after it is determined that human-centered design and development will be used, ISO 9241-210 (2010) specifies four linked human-centered design activities that must take place during the design of the system. These are shown in Table 2.1, together with example detail and the outputs they produce.

The next two subsections outline more detailed applications of HCD processes in mining that are compatible with ISO 9241-210. Work from the medical device design domain is then introduced to illustrate a potential future direction for the minerals industry.

TABLE 2.1
Human-Centered Design Activities

Activities	Detail	Outputs
1. Understand and specify the context of use	The characteristics of the users, tasks and organizational, technical, and physical environment define the context in which the system is used.	• Context of use description (e.g., user characteristics, tasks and goals, use environment)
2. Specify user requirements	User requirements provide the basis for the design and evaluation of systems to meet user needs. This includes user interface knowledge.	• Context of use specification • User needs description and requirements specification
3. Produce design solutions to meet these requirements	Potential design solutions produced based on the context of use description, the state of the art in the domain, design guidelines, and the knowledge of the design team.	• User interaction specification • User interface specification • Implemented user interface

4. Evaluate the designs against requirements	User-centered evaluation is a required activity at all HCD stages. Two widely used approaches are
	  • User-based testing, and
	  • Inspection-based evaluation by usability guidelines.
		• Evaluation results
		• Conformance test results
		• Long-term monitoring results

Source: ISO 9241, Ergonomics of human-system interaction—Part 210: Human-centered design for interactive systems, www.iso.org/iso/catalogue_detail.htm?csnumber=52075, 2010.

2.3.3   OTHER ISO DOCUMENTS TO SUPPORT MINING HCD

The title of ISO 12100 (2010) is: *Safety of machinery—General principles for design—Risk assessment and risk reduction.* It is not specifically about HCD, but it specifies a broad risk-based approach in which human-centered design activities could take place. The standard specifies basic terminology, principles, and a method for achieving safety in the design of machinery. It specifies principles of risk assessment and risk reduction to help designers in achieving this objective. These principles are based on knowledge and experience of the design, use, incidents, accidents, and risks associated with machinery. Procedures are described for identifying hazards and estimating and evaluating risks during relevant phases of the machine life cycle, and for the elimination of hazards or the provision of sufficient risk reduction.

2.3.4   EMESRT EDEEP

The Earth Moving Equipment Safety Round Table (EMESRT.org) is a global initiative of mining companies. EMESRT engages with key mining industry OEMs to advance the design of the equipment to improve safe operability and maintainability beyond standards such as the already-mentioned ISO 12100. EMESRT has introduced a task-based design evaluation process known as the EMESRT Design Evaluation for Equipment Procurement—EDEEP (n.d.). The purpose of developing EDEEP is to provide both OEMs and EMESRT members with a means to effectively identify the degree to which a piece of equipment meets the intent of the EMESRT design philosophy information.

EDEEP is a document designed to allow OEMs to provide information to equipment purchasers, describing the design controls implemented directed towards minimizing and mitigating risk within maintenance and operability tasks. The EDEEP document is made up of key sections directing the user towards a final document to be supplied to the purchaser for evaluation.

- Critical task identification information
- Design philosophy reference information
- Task-based risk assessment document information
- Design feature information from the Task-based risk assessment

In Chapter 3 (examples of successful mining HCD initiatives), an example of a major equipment manufacturer successfully using a modified EDEEP process is given.

2.3.5   MEDICAL DEVICE DESIGN

2.3.5.1   Applying Human Factors and Usability Engineering to Medical Devices

As an example of an emerging HCD process in a non-mining domain, the US Food and Drug Administration (FDA) has long stressed the importance of employing human factors/human-centered design techniques in understanding and controlling use error risks in medical device design and development. Their 2016 report

"Applying Human Factors and Usability Engineering to Medical Devices" includes the following (Food and Drug Administration, 2016a):

- Device users, use environments, and user interfaces
- Preliminary analyses and evaluations (e.g., task analysis or simulated use testing)
- Elimination or reduction of use-related hazards
- Human factors validation testing
- Documentation

The medical device design process is therefore comparable to the mining HCD processes outlined in this book. This includes understanding users and use environments, appropriate interface design, and user-centered validation testing.

The guidance also contains a summary reporting template to provide information related to medical device use safety. The report aims to show the key human factors considerations, issues, resolutions, and conclusions. The likely information that should be addressed in this report includes (Food and Drug Administration, 2016a):

1. **Conclusion**: The <device/equipment/technology> has been found to be safe and effective for the intended users, uses, and use environments.
2. **Description of intended device users, intended uses, use environments, and training** (including intended user population and context of use).
3. **Description of device user interface** (including a graphical representation of the device and its user interface).
4. **Summary of known use problems,** including those in similar/previous models, and design modifications implemented in response to user difficulties.
5. **Analysis of hazards and risks associated with the use of the device** (including potential use errors, and the potential and severity of harm that could result from each).
6. **Summary of preliminary analyses and evaluations** (including evaluation methods, key results, and design modifications implemented in response).
7. **Description and categorization of critical tasks** (including how critical tasks were identified and use scenarios for these critical tasks).
8. **Details of human factors validation testing** (including rationale test type used, test environment, training given, and test results)

The use of an HCD summary report such as this by mining technology/equipment manufacturers to show how they have addressed important human factors issues would be beneficial to help develop and deploy appropriately human-centered devices and technologies.

2.3.5.2 FDA Medical Devices: Design Controls

The Quality System Regulation's Design Controls (21CFR820.30) is the FDA's accepted process for medical device design (Food and Drug Administration, 2016b).

Human-centered design/human factors activities are essential at different stages of this FDA process. In Figure 2.1, the general process is summarized in a design

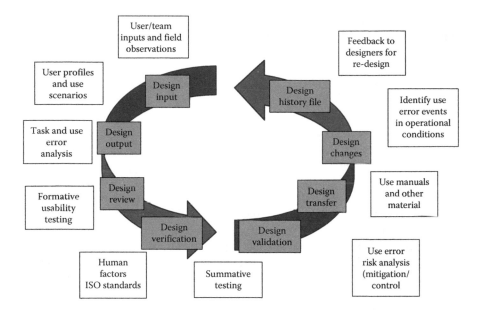

FIGURE 2.1 Example of a device design life cycle showing the FDA's accepted medical device design process and where human factors can support this. (Adapted from Human Centered Strategies, *Human Centered Design*, www.humancenteredstrategies.com/process. php, 2006.)

life cycle model: overlaid around it is where human factors/HCD activities can support this process (adapted from Human Centered Strategies, 2006).

In terms of HCD for mining equipment, the preceding work shows that human-centered inputs should occur across the design and development life cycle and at every step of the process. This ranges from the development of user profiles, task analysis, and observation in early design stages, through to testing, developing support materials, and giving feedback to designers in the latter stages.

2.4 HCD DATA AND TOOLS

As seen in the preceding section about processes, tools to collect basic data and to evaluate designs are an essential part of HCD. As Horberry et al. (2011) note, this includes in the testing process of designs (e.g., user acceptance trials) and also in providing input into different stages of the design process (e.g., regarding human sizes). The range of possible tools and data here is almost unlimited; however, the following two examples give an indication of what is available. Detailed tools are described later in the human-centered design educational material chapter.

2.4.1 HCD Tools

For Giacomin (2012), human-centered design tools can be classified based on their intended use. Tools can consist of facts about people such as anthropometric,

TABLE 2.2
HCD Tools for the Minerals Industry

Category	Example Data and Tools Available
Facts Regarding Mining Personnel	• Anthropometric data sets and models • Biomechanical data sets and models • Psychophysical data sets and models • Cognitive data sets and models
Capture of Design Requirements (Verbally based)	• Ethnographic interviews/questionnaires • Day-in-the-life analysis • Think aloud analysis • Be your customer/customer journey • Personas and scenarios (e.g., extreme users)
Capture of Design Requirement (Non Verbally based)	• Error analysis • Fly-on-the-wall observation • Customer shadowing • Facial coding analysis • Physiological measures
Simulation of Possible Future Operations	• Risk assessments • Participatory ergonomics • Focus groups • Codesign

Source: Giacomin, J., *What Is Human Centred Design?* P & D Design 2012 (10° Congresso Brasileiro de Pesquisa e Desenvolvimento em Design), São Luís (MA), Brazil, 2012.

biomechanical, cognitive, emotional, psychophysical, psychological, and sociological data and models. Giacomin's HCD data and tools that are relevant to mining are shown in Table 2.2.

2.4.2 Applicable ISO Document for Mining HCD

Focusing on user-centered evaluation, ISO 9241-210 gives two general methods. These are described as follows using mining as the application area:

1. **User-based testing**. Techniques that can be used to gather data from mine site evaluation include field reports, incident analysis, near-miss reports, log files, defect reports, user feedback, performance data (from mobile equipment, for example), satisfaction surveys, reports of health impacts, design improvements, user observation, requests for changes, and maintenance records.
2. **Inspection-based evaluation**. Inspection-based evaluation can be valuable and cost-effective and can complement user-based testing. It can be used to eliminate major issues before user testing and hence make user testing more cost-effective. As an example in mining, Horberry and Cooke (2014)

employed a usability audit checklist to review the draft interface of a prototype proximity detection system for mobile mining vehicles. Partly based on their inspection results, a revised interface was developed.

2.5   SUMMARY OF THE ESSENTIAL PRINCIPLES, PROCESSES, AND TOOLS FOR MINING HCD

A large amount of material has been presented in this chapter derived from mining, ISO standards and guidelines, and best practice human-centered design in another domain. Despite the wide variety of data sources used, there appears to be reasonable commonality in terms of the broad HCD principles, processes, and tools: these are summarized in Table 2.3.

As noted earlier, a human factors summary report by mining technology/ equipment manufacturers to show how they have addressed important human-centered design issues would be beneficial to help develop appropriate human-centered devices. Even without a regulatory requirement, this could be driven by larger mining companies requiring their equipment suppliers to provide an HCD report during procurement. Such an approach is contained within the "EDEEP"

TABLE 2.3

Analysis and Synthesis of the Key HCD Principles, Processes, and Tools Available for Mining Equipment and Technology Development

Principles	• The design is based on an explicit understanding of the user, their tasks, and the environment/use context.
	• Users and other stakeholders should be involved throughout design and development. Their needs, wants, and limitations are given explicit attention at each stage of the design process.
	• It fits the equipment, system, or interface to the user, not vice versa.
	• The design is iterative, evolutionary, and incremental.
	• It is driven by user-centered evaluation criteria (e.g., efficiency or usability) both during the design process and for the final end product.
	• A multidisciplinary design team is used, including HF/usability champions.
	• The design is integrated with the wider work system organization.
	• The HCD process must be customizable: capable of being adapted to different mine sites' conditions.
Processes	• Explore/investigate (understand the need and context of use and specify user requirements).
	• Produce/create design solutions based on point 1.
	• Evaluate the design (at all development stages).
	• Manage the process/feedback information to designers for the next iteration.
Tools	• To investigate/explore (e.g., observations and task analyses).
	• To provide input into stages of the design process (e.g., anthropometric data sets, human factors guidelines, or participatory design sessions).
	• As criteria in the evaluation process of designs (e.g., user acceptance trials, usability audit checklists, or long-term monitoring of the product/system).

process (EMESRT Design Evaluation for Equipment Procurement) (www.emesrt. org/emesrt-design-evaluation-for-eme-procurement/).

2.6   MINING HCD EXAMPLE: ADEQUATE UNDERGROUND LIGHTING

The work of Sammarco and colleagues at NIOSH PMRD, USA has shown that focusing on the needs and limitations of underground mine workers is vitally important in developing adequate lighting (e.g., Sammarco et al., 2009; Yenchek and Sammarco, 2010; Sammarco, n.d).

The issue. Providing adequate lighting to work safely is a challenge in an underground mine. An underground mine includes dust, confined spaces, and surfaces that reflect light poorly and have low visual contrasts. Lighting is critical for workers to see changes in the gradient of the ground, pinning and striking hazards, and slipping and tripping hazards. Age is also a significant factor: it becomes more difficult for older mine workers to see various hazards, and glare is more likely to impair performance.

NIOSH PMRD mine illumination research to improve worker safety. In a wide ranging program of work to improve a mine worker's ability to see hazards, topics such as cap lamps, machine-mounted lighting, glare, lighting maintenance, and light-emitting diode (LED) technology issues have been investigated. For example, new LED cap lamps were developed, first focusing on enhancing the color of light to improve the visual performance of older miners. Then, cap lamps were modified by changing the distribution of lighting so that floor and moving machinery hazards received more light to make them more visible. Work to evaluate them against current cap lamps indicated significant improvements for the oldest age group. Equally, the modified cap lamp enabled faster trip hazard detection and peripheral motion detection with no increase in glare compared to commercially available LED cap lamps.

Next steps. The latest research addresses the specific LED cap lamp needs for metal/nonmetal mining, where the visual environment and visual tasks are different compared to coal mining.

HCD conclusion. This NIOSH PMRD work employed all the essential aspects of human-centered design: focusing on the end user and their tasks, iterative design, and empirical measurement of performance. It studied the visual issues underground, reviewed injury data and examined worker visual performance (especially old workers), and then iteratively developed and tested prototype cap lamps. In this example, using HCD can help improve upon an existing design. In particular, the work found that improved cap lamps can be developed based on human-centered design-type processes.

3 Current Status of Mining HCD

3.1 OVERVIEW

This chapter presents a snapshot of mining HCD today. Although naturally limited to the time of writing, it presents overall trends in the minerals industry, and gives examples of recent mining human-centered design projects.

3.2 HOW MUCH MINING HCD HAS PREVIOUSLY BEEN UNDERTAKEN?

Although human-centered design can have many important benefits, until recently it was unclear exactly to what extent HCD actually had been used in the design and development of new mining technology and equipment. To help answer this, a database of new mining technologies and an analysis of how much human-centered design was used in their development were created.

The database was originally produced by Horberry and Lynas and published in *Ergonomics Australia* (Horberry and Lynas, 2012). A wide variety of data sources was used to create this database. In total, 108 entries were added, including entries relating to autonomous or semi-autonomous mining systems, proximity detection systems, and systems that support mine automation functions (such as control room management). A product description was given, including the technology used, the function of the system, and, where possible, the location it was being used (Horberry and Lynas, 2012).

The database and associated analysis showed that the global minerals industry was being changed by its increasing use of automated equipment and new technologies (Horberry and Lynas, 2012). At one end of the scale, this revolution is leveraging off-the shelf technologies to incrementally improve the control of various mining tasks, and at the other end, some initiatives are currently in progress to implement fully autonomous mining processes (such as automated drilling and blasting, or driverless trains). Technology developers are working within this space to provide stand-alone and integrated systems that provide solutions for the complexities arising from this massive change of focus within the industry.

However, one issue very clear from the database was the lack of focus on the operator: in other words, it showed the unpopularity of HCD in current mining technology development. Only approximately one-third of the database entries explicitly mention the impact these technologies make upon the operator; hence, it is likely that little use has been made of human-centered design methods for the majority of these technologies (Horberry and Lynas, 2012).

For mining HCD, it is therefore clear that many technologies have been developed purely from a technology-centered perspective that does not specifically consider the needs or abilities of the operator. Hence, human-centered design does not appear to be used for most new mining technologies. As noted throughout this book, experience from other industries shows that such human element issues must be considered or the technology will likely fail or at least work less than optimally. The widespread adoption of human-centered design processes and the involvement of operators at all stages of mining technology development and deployment are the key recommendations here.

3.3   MINING HCD: WHO IS GENERALLY DOING WHAT?

In addition to the database of new mining technologies, another approach to evaluate the current popularity of mining human-centered design is to review who is actually working in the mining HCD field. This includes workers in OEMs, academia, and in closely related fields.

For this book, this review focused on organizations/groups rather than individuals. Web searches, LinkedIn reviews, personal contacts, conference reviews, academic searches (e.g., Google Scholar), OEM websites/brochures, and email lists were used to identify who currently works in the mining HCD field.

In total, we found 32 organizations working in this area. These came from around the world, but mainly cluster around North America and Australia. The organizations identified cover the global minerals industry, and include OEMs, academia, institutes, industry groups, and government organizations. It also includes a few organizations working in broadly comparable fields (e.g., the automotive safety technologies work at Virginia Tech, USA). From this, it is notable that there are a moderate number of organizations, especially in academia, working in general human-centered design but not linked to mining, e.g.,

* Institute for Human Centered Design, Boston, USA
* Engineering Design Centre, University of Cambridge, UK
* College of Engineering, UW-Madison, USA
* Helsinki University of Technology, Finland
* Human-Centered Design Institute, Florida Institute of Technology, USA

Similarly, there are a large number of general mining groups. These often have a risk-management focus, not one specifically on human-centered design, e.g.,

* Cambourne School of Mines, UK
* Catholic University of the North, Chile
* Mining Engineering, Utah University, USA
* CRCMining: QLD, NSW, and WA, Australia
* Universities of Witts, Cape Town, and Pretoria, South Africa

Therefore, the **key finding here is that only a small number of organizations/ groups are specifically working in mining HCD**. These are generally restricted to the larger OEMs (e.g., Caterpillar and Sandvik), industry groups such as EMESRT, institutes/government-run groups (e.g., NIOSH PMRD USA), and academic groups (e.g., University of Queensland and Monash University Australia, or Laurentian University Canada). **The main conclusion is that mining human-centered design is a very small field.**

3.3.1 MINING EXAMPLE: EVALUATING UNDERGROUND ROOF-BOLTER CONTROLS

As an example of who is working in mining human-centered design, the research of Lisa Steiner (NIOSH PMRD, USA) (2014) and Robin Burgess-Limerick (University of Queensland, Australia) (Steiner and Burgess-Limerick, 2013) is presented. It gives a flavor of the type of mining HCD work undertaken in a joint American-Australian project.

The Issue. The interface between operators and mining equipment is critical to safe and effective operation. The types of injuries related to roof-bolting equipment are often crushing or striking accidents, potentially due to design deficiencies with roof-bolting machine controls.

NIOSH PMRD roof-bolting equipment research. Steiner and colleagues in USA and Australia conducted tests with roof-bolting machine control design to help limit future injuries. This operator-centered work began by an analysis of narratives describing injuries involving such equipment. This analysis revealed that some unintended movements occur when the wrong control is operated (a selection error) or when a control is operated in the wrong direction (a direction error). A series of experiments were then conducted in America and Australia (Figure 3.1): first, in virtual reality analogue; then via a physical simulation of a roof-bolting machine, and then utilizing a modified roof-bolting machine operated by experienced operators.

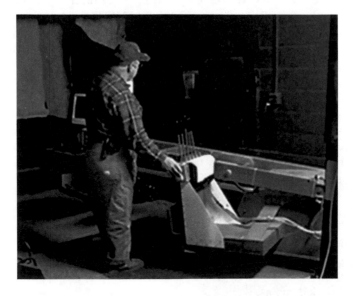

FIGURE 3.1   Experiments with bolter controls.

Three broad outcomes were produced by the experiments:

1. Ensuring that the design of equipment controls maintains compatible directional control-response relationships reduces the probability of direction errors.
2. Arbitrary shape-coding and/or length coding may reduce the probability of selection errors in some situations, although further research is warranted.
3. The implementation of a visual feedback system has potential to mitigate the consequences of unintended bolter movements.

Next steps. The next step in completing this research program is to undertake field trials of selected equipment modifications to determine their usefulness and practicality. Providing bolter-control design guidelines and principles to the mining industry is the overall goal.

HCD Conclusion. In terms of HCD, this research mainly examined the testing and evaluation aspects. The work is human-centered by its focus upon the operator: first, seeing what went wrong in the past (through injury narratives) and then by a series of experiments. It shows that roof-bolting machine controls can be redesigned to be better suited to human use. Although they focused on roof-bolting machines, these evaluation methods may be applied to the design of other mining equipment. This example also shows that international human-centered design research (in this case, between USA and Australia) can be beneficial.

3.4 BARRIERS FOR HCD

In this section, we review possible barriers and obstacles for HCD of mining equipment and new technologies. We begin by outlining key recent work about human-centered design barriers. We review legislative and MSHA USA approval processes for new technologies. Finally, we provide an analysis of the key barriers emerging: these are described in four broad categories: "the nature of mining," "the nature of humans," "design practice," and "selling HCD."

In this section, we conclude that perhaps the greatest impact and cost-benefit can be achieved through work on the fourth category, "selling HCD." Better and more accessible HCD methods/human factors information may help. Equally, case studies of successful human-centered design and developing HCD educational material would be highly beneficial. In this vein, this book, and especially the educational materials presented in the next chapter, should be of considerable assistance here.

3.4.1 HCD BARRIERS PREVIOUSLY FOUND

In this section, we summarize the human-centered design barriers identified in two previously published works. The first of these comes from mining (Horberry et al., 2011), and the second is a more general HCD summary that is applicable to many high-hazard work domains.

3.4.1.1 Human Factors for the Design, Operation, and Maintenance of Mining Equipment

In 2011, Horberry, Burgess-Limerick, and Steiner published the book *Human Factors for the Design, Operation, and Maintenance of Mining Equipment*. Because the focus was about both mining equipment and new technologies, the barriers identified and reviewed in that recent book are highly relevant here. Table 3.1 shows these barriers: the first column lists the general barrier/obstacle, and the second column provides additional details/explanations. It shows that HCD barriers can be caused by unseen/ unrecognized benefits of user involvement, human-centered design costs being seen as outweighing the benefits that would emerge, or the lack of knowledge about HCD.

3.4.1.2 Relevant HCD Barriers Identified Outside of Mining

Working outside of mining, Gulliksen et al. (2003) quote a broad survey that found the following barriers/obstacles to HCD. They are presented in order of how often they were reported.

1. Resource constraints
2. Resistance to user-centered design/usability
3. Lack of understanding/knowledge about what usability is
4. Better ways to communicate impact of work and results
5. Lack of trained usability/human factors engineers
6. Lack of early involvement by end users
7. Customers not asking for usability

TABLE 3.1
Mining HCD Barriers Identified by Horberry et al. (2011)

Type of Barrier	Additional Details/Explanation
Adding HCD/human factors to the design of a product is often seen as unnecessary	Reasons include mine site access issues, or benefits and costs of using a human-centered approach not clear
Humans differ in shape, size, ability, skill, and motivation	For example, in terms of mental processes like decision making or reaction time, motivation, or physical features such as a person's reach length
Lack of human factors knowledge/ overimputation	Designers get some of their information about the mine site user by attributing their own personal knowledge to others
Standardization does not work for all human-related equipment issues	In complex areas such as workload designers often have little assistance, and so need to rely more on trial and error or their own professional judgment
Competing priorities	When design compromises needed (e.g., lower price, higher usability), then immediate demands of pleasing client may take precedence
Professional pride of the designer and/or emotional investment in an idea	Designer reluctant to change something they have created, especially where HCD is not part of their core training
Perceived costs	HCD costs and benefits difficult to quantify. For example, "adding" usability may be seen as an additional cost

Gulliksen et al. (2003) offer no explicit countermeasures for these beyond applying rigorous HCD processes and advertising the success of these processes. To a large degree, all seven of these areas can apply to the mining/minerals industry. Indeed, given the conservative nature of the minerals industry compared to other domains such as the computer industry, it is likely that these factors will be of equal or greater importance in mining.

3.4.2   Legislative and MSHA Approval Processes for New Technologies

For new mining technologies, the regulatory approvals process can be a potential barrier. The example used here is the legislative and MSHA process specifically in the USA, but much of this is also applicable to other developed countries such as Australia, Canada, and most states from the European Union.

The US legislative and MSHA process for new mining technologies can be onerous: this is despite, for example, the Mine Improvement and New Emergency Response Act of 2006 (MINER Act) noting the need for technologies to be developed that can enhance self-escape from underground mines. Similarly, the National Research Council publication (2013) "Improving self-escape for underground coal mines" recommended that technologies enhancing self-escape should be developed using human-centered design principles. However, the report noted:

> The current technology regulatory and approval process in the United States appears to be a deterrent to rapid technology innovation and access to global markets, which hampers the commercial viability of innovation. (National Research Council, 2013, P6).

MSHA's mandates after a disaster require mines to implement solutions, often new technologies, in a short time period. Manufacturers must push their technologies through the MSHA approval process for them to be permissible in the mines. All installed technology is required to be intrinsically safe and/or housed in explosion-proof enclosures. This requirement adds to the complexity and often bulkiness of the design and, in addition, the approval process is lengthy and is unique to each design. Therefore, because the MSHA process takes a significant amount of time unless the technology is built upon previously approved platforms, the research, development, and design process is further shortened by this approval and certification process. The approval and certification process could be a deterrent to adopting an appropriate human-centered design approach due to its short time frame and specific design limitations. Many smaller companies have a difficult time entering the market and larger firms have little incentive due to the relatively small size of the mining market.

It is contended here that HCD guidance and tools for developing new technologies, along with a better understanding and the efficient execution of the MSHA design requirements, help to build better products for the industry and entice more manufacturers to participate.

3.4.3 ANALYSIS AND CONCLUSIONS

The preceding subsections examined different facets of the potential barriers to HCD in mining and related domains. From this, we identified four broad categories of human-centered design barriers for minerals industry equipment/technologies: these are presented in Table 3.2.

In terms of addressing the barriers in these four categories, the changing features of numbers 1 and 3 ("the nature of mining" and "design practice," respectively) are probably the most challenging to address directly. However, as will be seen later, they both can be addressed indirectly by means of improvements to the fourth category of "selling HCD."

The second category ("the nature of humans") is perhaps the most straightforward to address because the issue of human variability is a central tenet of human-centered design—and has been addressed well in other domains where HCD is better progressed (e.g., the healthcare or automotive industries). For example, issues such as physical variability in user groups can sometimes be accommodated by means of equipment adjustments, and diverse populations can be successfully designed for. Indeed, it can be argued that the diverse nature of user populations, often very different from those of the designer, is not really a barrier, but more of a key reason why HCD is necessary to help deliver mining equipment that fits the user.

However, it is argued here that perhaps the greatest impact can be achieved through work on the fourth category: "selling HCD." As noted throughout this book, better and more accessible HCD methods/human factors information may help. Equally, case studies of successful human-centered design and developing HCD educational material would be highly beneficial.

Changing the prevailing design mindset of an industry appears daunting, but as other domains (e.g., automotive) have found, there is a strong message to be conveyed concerning the benefits of HCD. For example, using a human-centered design process can minimize the need for initial and ongoing operator training due to the intuitive design of the equipment. Furthermore, in the minerals industry, a series of extensive interviews with mine site managers, OEMs, regulators, and new mining technology end users by Lynas and Horberry (2011a,b) found that: **all interviewees firmly agreed that unless human-centered considerations are deliberately part of the development process, then the new technology would most likely fail or not work optimally.** Thus, it appears the motivation to embrace human-centered issues does exist in mining, and providing examples of human-centered design successes and usable HCD educational material could be constructive.

3.5   EXAMPLES OF SUCCESSFUL HUMAN-CENTERED DESIGN INITIATIVES

The previous section highlighted the importance of providing examples of HCD successes; this section begins by outlining examples of such initiatives. For each example, we open by giving a short background section to introduce the issue, and then we describe the types of human-centered design activities undertaken and illustrate the outcomes.

TABLE 3.2

Analysis Results: Four Categories of HCD Barriers and Examples for Each Category

1. The nature of mining	A very conservative and risk averse industry
	Technology-centered design currently dominates
	Mine site reluctance to be the first to deploy unproven technology
	Legislative, intrinsic safety requirements and MSHA approval issues for new technologies
	Product liability lawyers reluctant to allow risk assessment data to be communicated
	Mine site access difficulties for designers
	Financial constraints by mining companies and OEMs
	Lack of mining customers asking for HCD during equipment/technology procurement
	Technology slow to be developed, sometimes longer than working life of the mine
	Uncertainly about future mines: how to get there and what equipment is needed
2. The nature of humans	Variable in shape, size, ability and motivation
	Diverse populations being designed for—often very different from the designer's own
	A diverse range of stakeholders—not just the end user, but also maintenance staff etc.
	Skill gaps for operating and maintaining new mining technologies/automation
3. Design practice	HCD can be seen as unnecessary by designers
	Designers unwilling to change—HCD not in their core training—use intuition instead
	For traditional equipment (e.g., trucks), new versions are often only small tweaks
	Other competing priorities (e.g., cost). Especially for smaller technology developers
	Lack of trained HCD or human factors professionals in mining companies and OEMs
	The MSHA approval process often results in little time being available for human-centered design processes
4. Selling HCD	Few case studies of HCD to "sell the HCD vision"
	Few cost benefit analyses—adding HCD seen as a cost
	Lack of easily accessible HCD guidance outside of ISO 9241-210 and Horberry, Burgess-Limerick and Steiner book
	Lack of understanding of (or even resistance to) HCD or HF by mining customers
	Lack of HCD communication within businesses or with customers
	HCD not consistent with mining "technology push"
	No single priority area to focus on (e.g., underground coal mining, or haul truck design)
	No obvious funding source to drive global HCD research and development in mining
	Lack of early involvement—HCD often only brought in when the design is largely fixed

Together, these examples show that HCD can be applied to different phases of the asset life cycle—for example, revising an existing product or helping to create a new technology. Also, they reveal that a wide range of methods can be used: including task analysis, team-based risk assessment, and user trials. The key feature that binds all these positive initiatives together is a ceaseless focus on users and their tasks throughout the iterative design process. **In conclusion, the examples show that human-centered design can be a highly successful and effective process.**

3.5.1 ELECTRIC ROPE-SHOVEL TECHNOLOGY

This example is adapted from Cloete, S. and Horberry, T. (2014).

3.5.1.1 Background

Electric rope shovels are large mobile excavators used in surface mining operations. They are controlled by a single operator from an internal cabin. In coal mining, the role of the shovel is to remove topsoil and overburden from an underlying coal seam. Material excavated from the face is loaded into haul trucks for removal. Shovels have generally become larger and more reliable; however, the methods of operation have remained essentially the same for over 60 years (Figure 3.2).

New technologies are being developed for shovel control (Onal et al., 2013). The research by Cloete and Horberry (2014) explored the HCD issues with two prototype technologies developed by CRC Mining and P&H/Joy Global: a semi-automated load-assistance system (AutoLoad) and a collision-avoidance system (TruckShield).

- **AutoLoad** is a semi-automated load-assistance protocol, intended to take control of parts of the shovel's operation cycle in which the moving payload is in close proximity to a haul truck. After the operator has completed digging, AutoLoad swings the loaded dipper over a waiting haul truck, deposits the load, and returns the dipper to a neutral predigging position.

FIGURE 3.2 Example of a shovel.

- **TruckShield** is a collision-avoidance system, designed to prevent metal-to-metal contact between the shovel bucket and the haul truck. It works by predicting the trajectory of the dipper and rapidly activating the hoist control if a collision is predicted.

Manual control can be resumed at any time during an AutoLoad cycle by depressing a "deadman" pedal; however, TruckShield, once it actively avoids a collision, does not allow operator veto.

The aim of the project reported by Cloete and Horberry (2014) was to ensure the systems-development process integrated the needs and capabilities of human operators by applying HCD techniques and principles to the evaluation and iterative design of the technologies for electric rope shovels.

3.5.1.2 What Types of HCD Activities Were Undertaken?

We conducted an operator-centered gap analysis of manual shovel operation (i.e., without operator assistance technology). Following observations, operator interviews, and technical data reviews, we developed a comprehensive hierarchical task analysis for shovel operation.

We then used the human reliability technique "HEART" to examine where the shovel operation task could fail from a human element perspective, and so where there was most need for technology to support the operator. The resultant needs analysis was then compared to the capabilities of the technologies under development.

3.5.1.3 Outcomes

The application of the HEART human-reliability process revealed task difficulties at several points of the digging operation cycle. It found that the prominent error producing conditions for the swing and dump subtasks would be effectively removed with AutoLoad and TruckShield working in concert. The outcomes supported the prototype technologies being developed: they indicated that the systems offer a good fit to the more difficult and error-prone aspects of shovel operation.

The approach employed proved useful for the design of assistive technology in mining. The work showed that human-centered design can work successfully with technology developers in mining even after an early prototype system had been developed. A detailed understanding of user requirements was highly beneficial, and also helped to identify possible future technologies such as a shared, real-time, in-cabin display for haul truck drivers and shovel operators depicting the position of the haul truck relative to the shovel.

3.5.2 EMESRT "EDEEP" AND SANDVIK "MAKE IT COUNT"

3.5.2.1 Background

Established in 2006, the Earth Moving Equipment Safety Round Table (EMESRT) is a global initiative involving major mining companies. EMESRT engages with mining industry original equipment manufacturers (OEMs) to advance the design of the equipment to improve safe operability and maintainability beyond standards. The purpose is to accelerate the development and adoption of leading practice designs to minimize the risk to health and safety through a process of OEM and contractor and user engagement (www.emesrt.org). As such, the end-user focus of EMESRT initiative clearly has much in common with mining HCD.

As one tool for OEM engagement, EMESRT introduced a task-based design-evaluation process known as the EMESRT Design Evaluation for Equipment Procurement (EDEEP). EDEEP provides both OEMs and EMESRT members with a means to effectively identify the degree to which a piece of equipment meets the intent of the EMESRT design philosophy (DP) information—this DP information includes best-practice human factors recommendations for operator workstation design (Burgess-Limerick et al., 2012).

3.5.2.2   What Types of HCD Activities Were Undertaken?

EDEEP is a document to allow OEMs to provide information to the purchasers of EME, demonstrating their actions towards minimizing and mitigating risk within maintenance and operability tasks through the use of design controls (www.emesrt. org/emesrt-design-evaluation-for-eme-procurement/). This includes the following to be supplied to the purchaser for evaluation: Critical task identification information, design philosophy reference information, task-based risk assessment information, and design feature information from the task-based risk assessment.

The global equipment manufacturer Sandvik recently made extensive use of the EDEEP process, partially modified in order to best suit the company's ways of working and offering product (Wester and Burgess-Limerick, 2015). The objective of the Sandvik EDEEP process was to look at the design of a piece of equipment and evaluate how potential safety risks could be mitigated (http://sandvikmakeitcount.com/safety/sandvik-edeep-process). Areas considered included operator visibility, ease of maintenance and operation, and equipment access.

Sandvik reported four cases using the EDEEP process. One of these is presented here. This involved conducting the process with a prototype of one of their new products, the DD422i underground drill rig, before the model was released to the wider market (http://sandvikmakeitcount.com/safety/sandvik-edeep-process/dd422i-finland).

Part 1 of the Sandvik EDEEP process involved Sandvik product-safety and product-design engineers identifying and observing the priority and critical tasks that needed to be addressed in relation to machine operation and maintenance. The critical tasks were filmed underground in the Sandvik Tampere factory's test mine whilst product specialists operated the drill rig.

Part 2 of the Sandvik EDEEP process was conducted at Agnico-Eagle Finland OY's site based in Kittilä, where both mine personnel and Sandvik representatives worked together to carry out a task-based risk assessment on the DD422i. As part of the overall risk assessment, Sandvik showed the customer films of Sandvik personnel operating the drill rig and asked the customer whether the customer performed the operating tasks as per the Sandvik operator and as instructed in the machine manual. This exercise gave both the customer and Sandvik the opportunity to discuss existing and potential safeguards for hazards that may arise.

Part 3 of the Sandvik EDEEP process was the creation of the Sales Design Information (SDI), which is the resulting document of the Sandvik EDEEP process. For the DD422i, the product-safety personnel of product area underground drilling created the SDI. This document was a report of the tasks conducted and the design features evaluated: it was created for equipment procurement/sales to mine sites.

3.5.2.3 Outcomes

The outcomes of the Sandvik-adapted EDEEP were that both the OEM and mine site end users judged the process as "a beneficial way of creating direct contact between those responsible for product safety and product development and crucially the people at the customer site responsible for operating and maintaining the equipment" (http://sandvikmakeitcount.com/safety/sandvik-edeep-process/dd422i-finland).

They noted that the EDEEP process allowed:

- Priority operational and maintenance tasks to be identified,
- Task flow charts to be created—based on how tasks are actually performed at a mine site,
- Risks to be identified and solutions for these to be developed in task-based risk assessments,
- Designs to be evaluated, and
- Documentation to be produced to show how the OEM has addressed the risks identified for particular equipment designs.

The lesson here for human-centered design is that end-user input is vital (Wester and Burgess-Limerick, 2015). Bringing together mine-site personnel and OEM designers through a task-based design evaluation process such as EDEEP can be highly effective to review designs before they are released, as in this example, or to produce fit-for-purpose retrofits to minimize and reduce risks when operating and maintaining existing machines (e.g., concerning ease of equipment access).

3.5.3 ONGOING NIOSH PMRD EXAMPLES

In addition to the preceding examples, the following ongoing projects/investigations in NIOSH PMRD present three examples of HCD at various stages. These are diverse not only in their approaches but also in their varying levels of technology and stages of completeness. They also support the potential development of a human-centered design roadmap to enhance design processes at NIOSH PMRD and in the wider mining industry.

The three examples described are intended to illustrate the criteria that need to be considered at any stage of design or redesign. The approach would preferably begin at the ideation/conceptual design stage to conserve resources and avoid poor outcomes. The three cases show that HCD can be practiced at all levels of the hierarchy of controls; however, it is most effective as the means to eliminate hazards. Finally, they show that human-centered design should become the way activities are ultimately done in mining: designing with humans in mind.

3.5.3.1   Continuous Miner Operator Positioning

Background: The mining industry and NIOSH PMRD have developed several technologies to address the positioning of an operator running a continuous-mining machine underground. Proximity detection is a means to warn operators when they are too close to the machine and to shut down the machine, or certain functions of the machine, when the operator is dangerously close. There has also been a good amount of ongoing work on the influence of visual cues and production on operator positioning. This information gives reasons as to why operators stand where they stand—in order, for example, to see certain features such as the sight lines, cutting head, roof conditions, or power cables. These pieces of work have been done separately and not necessarily considered together. This overlay is needed to represent the integration of health and safety (and productivity) to help operators make good operating decisions.

Importance: If a continuous-mining machine operator were to ask where the best place to stand around the machine is during different cuts in the production plan, would a definitive answer be possible? Does she/he have to choose between health, safety, and productivity? The three areas of health, safety and productivity can often compete with each other; therefore a holistic human-centered design approach is required here.

Preliminary data: Initial data have been obtained: it may be possible to overlay the health and safety information to find out whether there is a place that does not have a trade-off between health, safety, and productivity. But that integration of health and safety parameters has not yet been accomplished. Also, the overlay may reveal several solutions that point to gaps and solutions to help operators make a dual choice for health and safety. These gaps would be considerations for future work in the area.

Benefits for mining HCD: Showing how technologies developed for an individual purpose without consideration for all related subsystems and risks can conflict with each other even though the goal to reduce accidents and injuries is common.

Benefits for NIOSH PMRD research: Operators of continuous-mining machine use and are given information: sometimes conflicting information. They need to weigh health, safety, and productivity concerns when deciding where to stand to operate dangerous equipment and to expose themselves to other moving equipment. Giving operators specific information and educating them on the issues with their decisions helps make them more aware of the situation. Then they can make decisions that include productivity, health, and safety. This information may also help manufacturers of proximity devices to fine-tune their systems to meet the mine company's and the operator's goals. For example, it is difficult to worry about cumulative health effects that only become apparent years in the future when adverse safety consequences are more immediate.

3.5.3.2   Narrow-Vein Roof-Bolting Machine

Background: Jackleg drills have been used as standard method for drilling and bolting in narrow vein openings in some US mines. This operation is physically demanding and dangerous for roof bolters. In 2011, JH Fletcher & Co began development and testing of a narrow-vein roof-bolting machine specifically to replace jackleg drilling. The initial and primary goal of the machine design was to improve safety of the roof-bolting process in narrow-vein mining. However, to be accepted and effective, this machine had to be at least as efficient as the jackleg is now, provide improved safety and reduced physical effort, as well as require low/no-maintenance during the mining cycle. Fletcher asked OMSHR NIOSH to provide information and research into the design of the prototype and future prototypes as the design progresses. Task analyses with jacklegs as well as with the prototype are needed to bring all the issues to light and assess risk.

Importance: Jackleg drilling has been a hazardous and physically demanding job for decades.

An analysis of accidents, injuries, and illnesses reported by mine operators and contractors on MSHA Form 7000-1 from 2000 to September 2015 revealed that "Jackleg" was contained in 448 narratives during the period. Of the injuries involving jackleg drills, 90% occurred at metal mines. The majority of injuries associated with jackleg drills were relatively serious, resulting in days lost.

The most common mechanisms of injury were overexertion and slips and trips associated with manually handling the jackleg drills. This is perhaps not surprising given that the drills weigh around 40 kg. Use of the drills under an unsupported roof also exposes operators to risks of being hit by falling rock, as well as being struck by the drill itself. Injuries associated with drilling jumbos were less likely to involve days lost, and the most common injury mechanisms were associated with access or egress to the equipment, being struck by a range of different objects, and overexertion associated with manual tasks.

A number of health risks, including exposure to vibration, noise, and dust, are also associated with the use of jackleg drills in particular, although the cumulative nature of the adverse consequences mitigates against their presence in injury reports. For example, a field investigation of vibrations associated with jackleg drill operation revealed high levels (Keith and Brammer, 1994); a laboratory investigation of the noise exposure associated with jackleg drill operation (Camargo et al., 2010) suggested noise levels associated with drilling into granite reached 123 dB (A); and dust measurements from jackleg drillers provided a mean concentration of respirable silica of 69.7 $\mu g/m^3$ (Weeks and Rose, 2006).

In this case, it is entirely possible to implement a human-centered design approach enabling involvement at the ideation stage of design, where NIOSH PMRD has had input, and throughout the design stages, prototypes, and final design. This is a rare opportunity, as equipment development in mining is often built on another piece of equipment's proven design. Also, Fletcher Mining is willing to follow this process as an example of the correct process for design.

Preliminary data: The first design of the narrow-vein roof-bolting machine was developed with information from previous NIOSH PMRD studies and guidelines provided by NIOSH PMRD, as requested. The usability and productivity of this machine versus the jackleg will be assessed through a specifically designed task analyses and observations techniques used in an HCD approach at the test mines.

Benefit for mining HCD: There is an opportunity to conduct task analyses (with risk assessments) of all aspects of human use: dust, noise, and physical requirements, as well as to perform comparative analyses of the jackleg drill and narrow-vein bolter from these and other productivity criteria. This example has the potential to be a comprehensive "start-to-finish" human-centered design process.

Benefits for NIOSH PMRD: NIOSH PMRD's integral involvement in the method and analyses of an HCD approach will be a great model for future equipment design and will show the benefits scientifically of a human-centered design approach. No other equipment or process has followed this approach in the past. Also, this example supports the PtD Initiative in the US in terms of helping to eliminate hazards from equipment-related tasks.

3.5.3.3 Illumination

Background: In the past, the lighting requirements are determined for each piece of equipment by a process that considers the machine in only one position, with the appendages cocooned or in a neutral position. This is rarely, if ever, the way the machine is used and does not consider the user or the users' tasks or risks while performing those tasks. MSHA agrees that much work is needed on lighting. MSHA and Fletcher Mining Equipment are interested in different types of lights and the use of lighting to reduce accidents and injuries.

Importance: Lighting is a problem in underground mining, but lighting can be used to reduce injuries and accidents by properly highlighting critical tasks and pathways while providing additional feedback and information to the operators. Illumination can be used to help operators make better decisions and improve their situational awareness—enhanced situational awareness is a positive outcome of a human-centered design approach.

Preliminary data: Currently, lighting data have been collected for walkthrough bolter machines. Lighting alternatives have been discussed with Fletcher and MSHA. The lighting intervention for directional control and machine movement has been studied and is poised for implementation. The placement and type of lighting have been determined and the lighting is ready for placement.

Benefits for mining HCD: This work will exemplify the application of HCD principles in a system that has been used for many years. The ability to add appropriate lighting will go a long way to reduce "slip, trip, and fall" incidents around equipment, show how lighting can be used as cognitive feedback for proper machine-control movement, and provide warning systems for people approaching the area. The human-centered design approach will address issues that were not considered appropriately during the design process, but now new lighting technology is affording solutions that were not possible before.

Benefit for NIOSH PMRD: The functional analyses coupled with risk analyses of the roof-bolting machine (and the continuous-mining machine in the future) will help determine the appropriate lighting for the tasks that operators perform. Functional analyses consider the positions the machine would be in if operators were operating the equipment while bolting, tramming from place to place, or advancing the machine in the cut. These analyses, along with the risk analyses, will provide accurate location, positioning, and type of lighting needed to safely and effectively perform the tasks.

3.6　MINING EXAMPLE: DEVELOPING SELF-ESCAPE TECHNOLOGY WITH HCD

In Chapter 1, we introduced the need for self-escape technology to be developed using HCD-style approaches. To support this, we include the following summary example of how effective self-rescue equipment could actually be designed and deployed using a human-centered process.

The mining equipment life cycle from Horberry et al. (2011) is shown as the inner circle in Figure 3.3. Shaded text boxes around the inner circle describe examples of HCD inputs that could assist at different stages in the development of human-centered self-rescue equipment.

As seen at the top of Figure 3.3, the use of a human-centered design summary report by mining technology/equipment manufacturers could be beneficial to demonstrate how they have addressed key HCD issues. This approach is similar to that employed with some mining equipment in Australia (e.g., as required in MDG-15) and by the USA FDA for HCD in medical devices. During self-rescuer procurement, a technology manufacturer could provide this report to mine site purchasers and regulators.

Overall, this brief example shows how HCD could generally be used to help develop self-rescue equipment that is effective, usable, and acceptable to operators. In the next two chapters, this overall approach is substantially expanded through human-centered design educational material and detailed case studies.

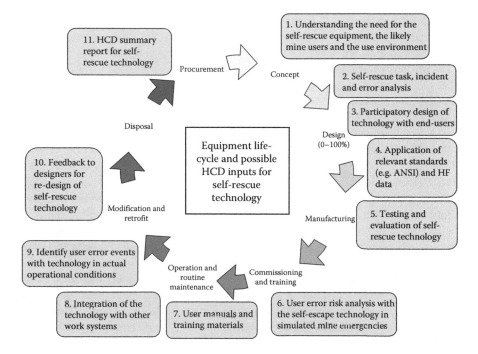

FIGURE 3.3 Example of developing effective self-escape technology using HCD-style approaches. (Adapted from Human Centered Strategies. Human centered design. www.humancenteredstrategies.com/process.php, 2006.)

4 HCD Educational Material

4.1 OVERVIEW

In this chapter, we present a "how to" educational guide for HCD, which breaks down human-centered design into 11 different activities. These 11 activities are undertaken at different points in the overall life cycle of designing, deploying, using, modifying, disposing of, and procuring mining equipment. We provide examples to help describe the 11 different activities. We anticipate that this educational material will be of use to promote the greater use of HCD in projects with mining equipment and new technology in the minerals industry.

4.2 HOW TO UNDERTAKE HCD

As noted earlier in this book, there is no single prescribed method of undertaking HCD; however, the approach presented here is consistent with the most relevant ISO standard (ISO 9241-210, 2010). It is also consistent with leading design approaches used in product design. For example, one product design approach focuses on four phases: *explore, create, evaluate,* and *manage*: the *explore* phase determines the needs; the *create* phase generates ideas to address how the needs can be met; the *evaluate* phase judges and tests the design concepts to determine how well the needs have been met; and the *manage* phase works across the whole process, reviewing the evidence to decide what should be done next (EDC, 2017).

The approach we present incorporates HCD across the full life cycle of equipment: from original concept through to eventual disposal. It also expands on the HCD self-escape technology example that was presented at the end of the previous chapter

4.2.1 THE EQUIPMENT LIFE-CYCLE

Figure 4.1 shows a basic life cycle for mining equipment.

This life cycle is broadly applicable to all mining equipment and technology; however, it is most pertinent for "traditional" equipment such as haul trucks or roof-bolting machines. Although shown as a circle, a key starting point can be the design "concept" stage: after this comes design, equipment manufacturing, equipment commissioning at a mine site (including training), operation and routine maintenance of the equipment, any modification or retrofit of it (e.g., "smart" technology), eventual disposal of the equipment (e.g., sale or scrap), and then purchasing/procurement of new equipment.

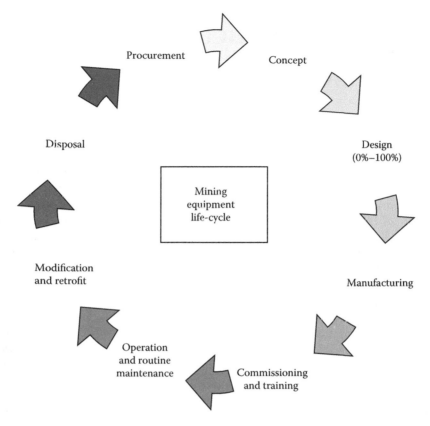

FIGURE 4.1   Life cycle of mining equipment.

4.2.2   HCD with the Equipment Lifecycle

Different human-centered design activities fit at different stages of the life cycle. Unsurprisingly, a large number of them occur at the concept/design phases, but HCD still has a valuable role to play later in the mining equipment life cycle.

In Figure 4.2, the basic mining equipment life cycle is shown in the inner circle. Around it are 11 shaded text boxes that describe where HCD inputs can assist at different stages in the development of human centered equipment and technology. As noted at the end of Chapter 3, a very similar approach has been successfully used for medical equipment, and Figure 4.2 is partly adapted from Human-Centered Strategies, 2006 (www.humancenteredstrategies.com/process. php).

However, it should also be stressed that design often needs many iterations, especially in the early stages, so the approach presented here should not be thought of as a linear process. Such iteration may also include wider issues such as refining and managing the overall scope of the design and building a business case to help demonstrate the technology/equipment's potential for profitability (EDC, 2017).

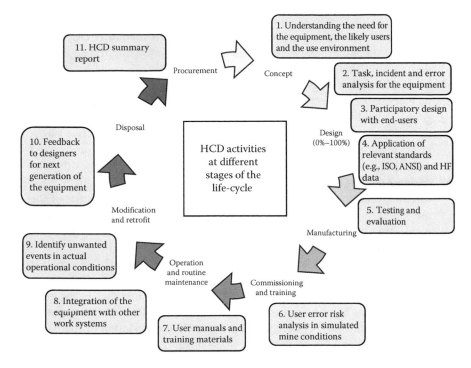

FIGURE 4.2   HCD activities at different stages of the mining equipment life cycle.

4.3   EDUCATIONAL GUIDE FOR THE 11 DIFFERENT HCD ACTIVITIES

4.3.1   UNDERSTANDING THE NEED FOR THE EQUIPMENT, THE LIKELY USERS, AND THE USE ENVIRONMENT

4.3.1.1   Purpose

The purpose of this step is to understand the need for the equipment from an end-user perspective: who will be using the equipment, what are their requirements, and what is the likely use context?

4.3.1.2   Rationale and Recommended Approach

A wide range of methods are available to gain a deeper understanding of the need for equipment/technology, the likely users, and the use context. For example, Stanton et al. (2013) note 107 human factors design and evaluation methods, and recommend the use of a comprehensive technique called cognitive work analysis to help identify the need for equipment. However, without extensive training, this technique is difficult and time consuming to apply. Instead, we recommend the following three practical actions:

1. **Observe users to determine what they really do, and consequently, what they really need**. Observing actual behavior is vital because people may struggle to express their real needs. Observations can supplement and

expand on interviews or questionnaires because people cannot imagine all the possible design alternatives to the current situation (EDC, 2017). Focusing on actual user behaviors helps avoid overloading the equipment with features that designers think the users want.

2. **Understand the likely users and specify user requirements**. This includes understanding the range of users, their characteristics, and their goals. It also includes getting an understanding of current and likely future users: for example, traditional drivers of haul trucks and prospective supervisors of automated haul trucks. One way to capture this is a requirements/needs list: a comprehensive list of the users and operational needs that the design solution should satisfy. The purposes of such lists are to provide links between the requirements of the design project and the needs of the users, and to enable prioritization of different needs (EDC, 2017). In other words, this list helps to identify fundamental *who?* and *what?* questions related to the proposed equipment or technology.

3. **Describe how and where the equipment would be used**. The context of use refers to all the circumstances in which interaction with the equipment/technology takes place: this includes physical factors such as the mine site lighting or how work is conducted in teams (EDC, 2017). One way to accomplish this is by creating a "use journey": a description of the likely context of use and the probable use environment (including organizational and physical environment aspects). It should consider purchase, commissioning, relocation, and modification, as well as "normal" use. This information can expand and enhance the earlier-mentioned requirements/needs list to also consider *where?*, *how?* and *why?* questions related to the proposed equipment or technology. This understanding can then feed into the next stages of the design process, and user testing can be repeated throughout the design process to ensure the finished product is safe, effective, and usable (Sharples et al., 2016).

4.3.1.3 Example: The Need for Shovel Assist Technology

Cloete and Horberry (2014) visited three Australian mines to undertake shovel observations and informal interviews. The aims were to understand actual operating practices and the operators' opinions of prototype load-assist technology. They observed a range of different operating conditions, face conditions, dig and loading patterns, operating styles, and levels of operational experience. Views of task difficulty, error potential, and technological interventions were identified with the operators: this included visibility with off-side loading. This led to an understanding of the human-error potential, and how a prototype technology could assist in supporting the most error-prone aspects of shovel loading.

4.3.1.4 Example: Accommodating Variability in User Anthropometry

People come in a wide variety of sizes and shapes. Designing to accommodate the range of potential users can pose a challenge. For example, underground coal-mining mobile equipment is frequently designed with severe height constraints, which in turn limits the cab height available, making the accommodation of tall users problematic.

Seat-height adjustability can assist, but the range is typically limited. In the end, the designer may need to specify the range of operator heights able to be accommodated by the design and communicate this to purchasers, thus constraining the likely users.

4.3.2 TASK, INCIDENT, AND ERROR ANALYSIS FOR THE EQUIPMENT/TECHNOLOGY

4.3.2.1 Purpose

The purpose of this step to understand the likely tasks with the equipment/technology and what has gone wrong (errors or incidents) in the past with similar equipment/technology.

4.3.2.2 Rationale and the Recommended Approach

Effective equipment or technologies cannot usually be developed without first understanding the problems experienced with similar systems in the past. From a HCD perspective, to understand the likely tasks and what has previously gone wrong, we recommend the following two activities.

- **Task analyses of equipment operation and maintenance**. Task analyses are essential to understand how equipment will be used, maintained, and sometimes even misused (Stanton et al., 2013). Task analysis identifies the subtasks required for the safe and efficient operation of the equipment. Where there are differences between the way a task is prescribed (i.e., the "official" version of managers or manuals) and the way that task is routinely conducted, this is valuable information for redesign (Simpson et al., 2009). Task analysis is also a key method for remote-control systems: for example, to understand the visual, auditory, and haptic cues a load-haul-dump operator uses when maneuvering the vehicle manually to identify if these are adequately addressed in a remote-controlled system.

  Task information may include video footage; review of existing documentation such as operating manuals, or safety reports; any previous task analyses; observations; and questioning of the people actually performing the task about what they do, how they do it, and what information they use. A widely-used approach represents the tasks in hierarchical diagrams (Kirwan and Ainsworth, 1992).

- **Incident and error analysis**. Analysis of incident, accident, or error data can help pinpoint equipment-design deficiencies (e.g., access and egress on bulldozers). For error analysis in mining, one approach is "human reliability analysis" using approaches such as "HEART" to estimate the error potential for different subtasks (Cloete and Horberry, 2014). HEART allows numeric probabilities to be associated with error occurrences. Originally designed for the nuclear and chemical industries, it is a structured and easy-to-understand technique. The key steps of the method are (Williams, 1986):
  - Determine the task or scenario to be analyzed.
  - Conduct a hierarchical task analysis for this task or scenario.

- Conduct a HEART screening process (a set of guidelines to identify the likely causes and sources of error for the task/scenario being analyzed).
- Classify task unreliability (that is, define a proposed nominal level of human unreliability for the task/scenario under review).
- Identify error producing conditions.
- Assess the proportion of effects of the error producing conditions.
- Identify and propose remedial measures for the errors identified.
- Document the full process.

It is not the place here to comprehensively describe the details of the HEART process, but good reference sources do exist. For example, see Stanton et al. (2013) for a full description of the HEART method, and Cloete and Horberry (2014) for an application of HEART for technology in the minerals industry.

4.3.2.3   Example: Mining Equipment Injury Data Analysis

Analysis of the narratives describing equipment related injuries reported to regulators has been utilized as a means of identifying opportunities to improve mining equipment design (Burgess-Limerick, 2005, 2011; Burgess-Limerick and Steiner, 2006, 2007; Cooke, 2015). For example, Figure 4.3 describes the activity being undertaken and the mechanism of injury for 593 injuries associated with underground coal roof-bolting machines in US mines (Burgess-Limerick and Steiner, 2007). The large number of "struck by falling rock" injuries while operating bolting machines pointed

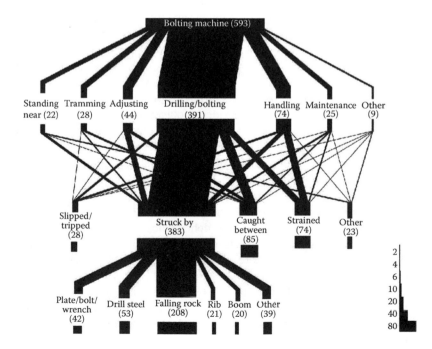

FIGURE 4.3   Activity being undertaken and injury mechanism for injuries associated with underground coal bolting machines at US coal mines.

to potential design changes, as well as highlighting the importance of installing roof mesh to prevent small rocks falling onto the operators of such equipment.

4.3.2.4 Example: Mining Fatality Investigation Analysis

Bow-tie analysis (sometimes called "cause-consequence" analysis) is a risk analysis and communication technique widely used in high hazard industries (e.g., aviation, chemical, petrochemical) and, more recently, in mining (Burgess-Limerick et al., 2014). The technique combines elements of fault-tree analysis and event-tree analysis. At the center of each bow-tie is an initiating event, the point in time when there is a loss of control of a hazard. The causes of the initiating event, and the potential consequences of the event are then determined, as are the control measures that can reduce the probability of the initiating event occurring (preventive controls), and the measures that can be taken to reduce the severity of the consequences of each initiating event (mitigating controls).

In an analysis of the reports of MSHA investigations undertaken into all fatal mine accidents occurring in the USA over a 10-year period (2005–2014) [NIOSH Contracts 200–2015-M-62391 and 200–2014-M-59063], bow-tie representations of each event were created as a means of identifying both existing and potential control measures that may have prevented the initiating event from occurring (preventive controls), or that may have reduced the severity of the consequences of the initiating event (mitigating controls). The focus was on technological controls rather than administrative controls such as procedures and training. The frequency with which different control technologies were included in the bow-tie representations, and the number of fatalities in which each may have potentially had a beneficial effect, is examined as one means of prioritizing potential technological control measures for further investigation.

The investigation reports for 451 fatal mining accidents occurring in the USA from 2005 to 2014 were retrieved. Single fatalities resulted from 433 of the incidents. Thirteen incidents involved dual fatalities and one involved a triple fatality. Four major incidents occurred during the period in which 5, 9, 12, and 29 fatalities occurred, respectively. The total number of fatalities over the 10-year period was 517. Bow-tie representations of each event were created as a means of identifying control technologies that may have prevented the initiating event from occurring or reduced the severity of the consequences of the event.

Based on the number of fatalities over the 10-year period in which each of the control technologies identified may have been beneficial, the priority technologies defined for further investigation were: interlocked pedestrian proximity detection on mobile equipment—incorporating proximity warning (46 fatalities); interlocked seat restraints on mobile plant, particularly trucks (39); video cameras on mobile plant (37); remote operation of mobile plant (37); noncontact methods of assessing underground strata condition (36); fire suppression on fixed and mobile plant (33); remote methane monitors in gob interlocked with longwall shearer (29 fatalities); methane extraction from longwall block in advance of mining (29); remote operation of longwall (29); active explosion barrier (29); stone dust monitoring (29); interlocked proximity detection on fixed plant (27); non–line-of-sight remote control of continuous-mining machine (26); usable SCSR/CABA (19); refuge chamber (19);

inertization of sealed areas (17); remote monitoring of sealed areas (17); live electrical warning devices (16); remote bolting (14); automatic brake testing at prestart (13); and outburst prediction (12).

Many of these control technologies are specific to underground coal mining, a consequence of nearly 40% of the fatalities occurring in this group of mines. The highest priority control measures for surface coal mines are interlocked seat restraints and the remote operation of mobile plant. These controls are also relevant for surface stone/sand/gravel sites; however, interlocked pedestrian detection for fixed plant such as crushers and conveyors is the highest priority for this group.

Many of the technologies identified are currently under development or investigation. For example, remote methane monitoring and noncontact methods of assessing strata condition are topics identified in calls for proposals by NIOSH. Stone dust measurement techniques have also been recently developed (Barone et al., 2015) and research is being conducted on alternatives to stone dusting.

Other technologies have been in development for some time. For example, considerable work has been undertaken towards automation and remote operation of longwall equipment and towards demonstrating the feasibility of active explosion barriers. The assessment of outburst risk has also been the subject of a recent review (Gray and Wood, 2015). Considerable work has also been undertaken in the area of interlocked pedestrian proximity detection for underground coal mining mobile equipment and this technology is commercially available for continuous-mining machines. Other technologies, such as video cameras for mobile equipment (infrared or visible spectrum), remote operation of dozers, and methane drainage, are also available; although the extent to which such technologies are deployed has not been systematically assessed. Advanced methane draining techniques using water jets are also under investigation, and it has been suggested that methane from in-seam gas drainage may be used to achieve inertization of a longwall gob (Claasen, 2011). Other technologies are less well developed, for example, interlocked pedestrian proximity detection for fixed equipment, and non–line-of-sight remote control for continuous-mining machines, although substantial efforts have been undertaken by NIOSH PMRD towards remotely supervised continuous mining machine operation in the past. Non–line of sight remote operation of continuous-mining machines has also been previously achieved for temporary use in outburst conditions.

Determining the priority of different control technologies also requires consideration of potential effectiveness. However, estimation of the likely effectiveness of any particular control technology in a given situation is not trivial. The potential effectiveness of control technologies, and their implementation success, will be dependent to a large degree on the effort undertaken during design and implementation to ensure that human capabilities and limitations are considered. For example, the installation of video cameras is a potential control technology to reduce the probability of loss of situation awareness caused by restricted visibility from mobile equipment. However, in any given situation, the probability of loss of situation awareness will not be reduced to zero. The effectiveness of the technology depends on its design with respect to human limitations and capabilities. The location of video displays within the cab will influence an operator's ability to assimilate and utilize the visual information. The technology is also subject to barrier decay mechanisms, such as the need to maintain the technology in good working order.

Another example is the potential for interlocked pedestrian proximity detection technology to be installed on mobile haulage such as shuttle cars, RAM cars, and Load-Haul-Dump in underground coal mines. The argument for pursuing such a technology is strong, in that such equipment was involved in 16 fatalities involving interactions with pedestrians in underground coal mines in the 10-year period examined here. (In contrast, only nine fatalities were associated with continuous-mining machine-pedestrian interactions in the same time period). However, while there may be technical issues to overcome, there are also substantial human factors issues that may prevent the effective deployment of the technology. For instance, no detection technology is perfect, and there will be false positives. The implementation of the control measure will fail if the false positive rate is not acceptable to equipment operators.

Consequently, investigation of the most acceptable false-positive rate to users, and what influences this rate, will be an important part of developing the technological control. Similarly, the installation of interlocked pedestrian proximity detection technology on mobile haulage has the potential to alter the behavior of both the drivers of mobile plant and pedestrians working in the vicinity of mobile plant in ways that are likely to decrease the effectiveness of the control measure. These behavioral changes also require consideration during the development of the control measure. Similar considerations apply to each of the other control technologies identified and require further exploration.

4.3.3 PARTICIPATORY DESIGN WITH END USERS

4.3.3.1 Purpose
The purpose of this step is to work with end users to iteratively design and evaluate new versions of equipment/technology.

4.3.3.2 Rationale and the Recommended Approach
The heart of any "participatory ergonomics" approach is the assumption that the end users and other stakeholders (e.g., maintainers or supervisors) are the "experts" and must be involved throughout the design life cycle if the equipment/technology is to be designed and deployed successfully. Working with end users and designers/manufacturers can help create possible solutions to meet the previously-identified user needs and requirements, as well as validate and potentially expand the list of user needs/requirements (Sharples et al., 2016). Participatory approaches can work well throughout the design life cycle, as manufacturers may build increasingly detailed mock-ups and prototypes of the technology or equipment: this includes simple desktop/paper-based mock-ups and computer drawings with only limited interaction through to complex, interactive simulations.

As noted earlier, evidence exists to demonstrate the effectiveness of such participatory approaches in mining (Burgess-Limerick et al., 2007; Torma-Krajewski et al., 2007). Compared to non–user-centered methods, the participatory approach may take longer and be more iterative, but it is more likely to lead to beneficial outcomes (Cantley et al., 2014; Rivilis et al., 2008).

There are a wide variety of participatory ergonomics methods, including "distributed participatory design," "scenario-based design," or "co-design," but two methods that have been used successfully in the minerals industry are

- **Decision-centered design.** The decision-centered design process adapted to mining (Horberry et al., 2011) involves five iterative steps: *preparation* (to gain an understanding of the domain, users, equipment and tasks; this includes identifying key tasks), *knowledge elicitation* (where task analysis methods are used to understand what are the key decisions made by operators when interacting with the equipment), *analysis and representation* (where data are decomposed into discrete elements in order to identify the key decision points), *application and design* (where knowledge about determining how to best support the identified decisions being made is translated into design concepts and prototypes) and *evaluation* (where the new designs are tested and the impact estimated). The participatory design phase in this process identifies the key decision points likely to occur when using the equipment/technology and uses them as a basis for interface design and skills training to best assist the operator.
- **Safety in design ergonomics.** The safety in design ergonomics process (also known as OMAT) usually involves seven stages. Stages 1–4 are usually performed in joint designer/end user workshops (Horberry et al., 2014).
  - Stage 0 defines the context and scope of the whole process (e.g., to review design deficiencies on an existing piece of mining equipment— such as bulldozer access and egress).
  - Stage 1 identifies the critical tasks; for mining equipment, a full list of maintenance and operational tasks performed may already exist. From these general task lists, critical tasks are prioritized.
  - Stage 2 involves decomposition of the high priority tasks identified in Stage 1.
  - Stage 3 involves identification of risks at each of the subtasks from Stage 2 (usually considering both likelihood and severity).
  - Stage 4 is the development of human-centered solutions for the risks previously identified in Stage 3. The input of both end users and designers is the key item here.
  - Stage 5 involves iterative development and evaluation of the Step 4 solutions by the designers/manufacturers.
  - Stage 6 involves managing the whole process and documenting it.

4.3.3.3 Example: Participatory Design of Moranbah North Longwall

In 2007, Joy Global designed a new longwall for Anglo Coal Australia's Moranbah North underground coal mine in central Queensland. The 1,750-tonne capacity powered roof supports were larger than any previous design, and this created a number of human-centered design challenges (Figure 4.4).

FIGURE 4.4   Joy Global longwall, including powered roof supports.

A team of operators and maintenance workers from the mine and an external ergonomist participated in the design with the manufacturer. The participatory process included visits to site by Joy designers to observe the previous longwall in operation at the time, numerous meetings, risk assessment activities, and design reviews culminating in a two-week visit by the team of operators and maintainers (and consultant ergonomist) to the Joy manufacturing facility in Wigan (UK) to inspect and modify the prototype supports.

One challenge associated with the size of the supports was facilitating the safe movement of miners along and between the front and rear walkways. Safe design features arising from the participatory design process included: extensive area lighting provided with magnetic base to allow movement as required by operators and maintainers; front walkway cover plates over push-rams to provide a continuous and even walkway (see Figure 4.5a through c); steps and collapsible railings for access to manual canopy controls used during maintenance; nonslip treatments on all likely stepping places, including valve covers; handles on legs to assist moving between front and rear walkways; and providing steps for access to the face conveyor.

4.3.4   Application of Relevant Standards (e.g., ISO, ANSI) and Human Factors Data

4.3.4.1   Purpose

The purpose of this step is to apply relevant standards, guidelines, and human factors data to the design of the equipment/technology.

4.3.4.2   Rationale and Recommended Approach

A large number of standards are relevant to mining equipment and new technology; these include ISO 9241-210 (2010), ISO 12100 (2010), ISO 6682 (2008), MDG 15 (2002), ANSI Z590.3 (2011) and European equipment safety directives. Equally, a huge number of human factors design guides exist, such as those by the Federal Aviation Administration or the US Department of Transportation. Similarly, larger amounts of data on user capabilities (e.g., vision, hearing, dexterity, and mobility for inclusive design) are emerging (e.g., EDC, 2017). One key reference source that covers much of this for the specific design of mining equipment is Horberry et al. (2011).

Generally, a minimum of five important areas should be considered for most mining equipment that involves human interaction: the first two of these (anthropometric

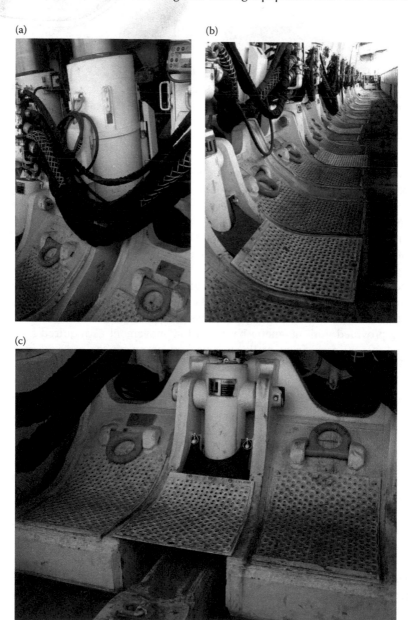

FIGURE 4.5　　(a–c) Walkway cover plates.

data and manual tasks) are topics to take into account, while the latter three are key aspects of the design that should subsequently be considered.

1. **Size, weight, and body shape: Anthropometric data sets and models**. Commonly available data are static one-dimensional ones such as height. Dynamic anthropometric data such as reach distances are potentially more

useful. Anthropometric data is available that may be suitable for mining equipment design. Data are available in software packages (such as PeopleSize), human factors textbooks devoted to anthropometrics (such as "Bodyspace"), AS4024.1704 (2006), or US Department of Defense Military Handbook Anthropometry of US Military Personnel.

2. **Manual tasks**. The equipment design process should include identifying potentially hazardous manual tasks involved with equipment and documenting the associated risks. Online database software and training is available to facilitate this process (www.ergoenterprises.com.au). Where key manual tasks risks exist, additional design control measures should be explored (ISO 12100, 2010), ideally using measures that are high up the "hierarchy of control" (for example, elimination controls).

3. **Work-station design**. Key issues here include clearance requirements (e.g., head or knee room); access/egress, and fall prevention; avoiding awkward postures; location and arrangement of workstation controls and displays (e.g., by temporal sequence of use, functional grouping, frequency of use, spacing of controls, location compatibility between controls and associated displays, and optimal display location) (Horberry et al., 2011).

4. **Control design**. Major issues include the choice of control type, resistance, control order, coding, and directional compatibility.

5. **Display design principles**. Key issues include to avoid requiring absolute judgements; to utilize redundancy; to use pictorial realism; to minimize the effort required to access information; to minimize the amount of information displayed (e.g., displaying only the most important information for decision making and only when needed), to have proximity compatibility; to use predictive aiding, and to ensure consistency in all the design.

4.3.4.3 Example: Roof-Bolting Machine Control Design

Some injuries associated with the operation of manual controls, such as those found in underground coal roof-bolting equipment, arise as a consequence of errors made in the use of the controls, including "selection errors" in which a control other than the one intended is operated, or "direction errors" in which the intended control is operated but in the incorrect direction. Figure 4.6 shows an example of bolter controls. A series of human factors investigations (Burgess-Limerick et al., 2010a,b; Steiner et al., 2013,2014) described the issues and proposed principles for the design of such manual controls, which were subsequently included in guidance materials provided by the NSW mine safety regulator (MDG35.1). Perceptual countermeasures have also been proposed to facilitate the recovery from such errors (Steiner et al., 2013). Incorporating these principles into subsequent equipment design is an example of human-centered design using human factors data.

4.3.4.4 Example: Dragline Seat Control Layout

A range of relevant standards may be consulted for guidance. For example, Figure 4.7 illustrates an assessment of the location of controls for a proposed dragline seat prior to its installation. Such a design could be compared against ISO 6682 (2008) "Earth-moving machinery—Zones of comfort and reach for controls."

FIGURE 4.6   Roof bolter control design.

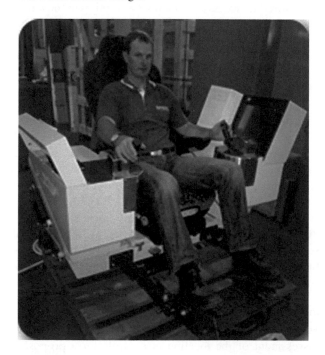

FIGURE 4.7   Assessment of the location of controls for a proposed dragline seat prior to its installation.

4.3.5   TESTING AND EVALUATION OF EQUIPMENT/TECHNOLOGY

4.3.5.1   Purpose

The purpose of this step is both to test the design as it progresses, and to evaluate the completed equipment/technology from a human-centered perspective.

4.3.5.2   Rationale and the Recommended Approach

Human-centered testing and evaluation is of two primary types: formative and summative evaluations (ErgoTMC, 2016). Formative evaluations are performed as the design progresses, iteratively feeding results back to the design to improve it. Summative evaluations are for completed designs of equipment/technology, before it is deployed at a mine site. For both types, actual end users, subject matter experts, and human factors staff can all participate. As an example, the UK Ministry of Defence uses a human factors assessment framework in three levels: rapid inspection and evaluation, field trials with representative end users, and controlled laboratory trials and simulations (Sharples et al., 2016). From a human-centered design perspective (ISO9241: 210, 2010; ErgoTMC, 2016; EDC, 2017), the following approaches are recommended:

1. **Formative evaluation.** The key steps here are: begin it early in the design process; assemble a team of five or more participants (including end users and designers); decide which design issues are to be reviewed and evaluated; review them; propose changes (if needed); get feedback on these from designers; document outcomes; and iteratively repeat the process as needed throughout design.
2. **Summative evaluation: inspection-based.** A key approach commonly used is a checklist evaluation performed by human factors evaluators to confirm that all intended functionality intended was correctly implemented. Steps include developing a list of criteria by which different aspects will be measured and evaluated, creating a checklist that covers them, using it to audit the completed design, and, where possible, supplementing this with performance-based testing where real users of the system perform real or simulated tasks.
3. **Summative evaluation: user-based**. User acceptability/acceptance and usability are key approaches here, often measured through structured surveys. Key steps include developing a survey instrument that asks focused questions about the acceptance/usability of the equipment, applying the survey with a sample of end users, and quantifying the results. These can be supplemented with relevant mine site operational data such as incident analysis, near-miss reports, defect reports, real user feedback, performance data (e.g., from mobile equipment), user observation, requests for changes, and maintenance records.

4.3.5.3   Example: Operator-Centered Evaluation of Mining Proximity Detections Systems

Cooke (2015) worked with a developer of a proximity detection system for underground hard-rock mining to evaluate the systems using an operator acceptance survey. He used an established method for measuring acceptance developed by

FIGURE 4.8   A prototype proximity detection system.

Van der Laan et al. (1997): a rating scale for nine acceptance questions (such as how "useful" the system was). He tested operator acceptance of both a prototype proximity detection system (Figure 4.8) and the final system. Operators found the final system much more acceptable than the older version. Cooke attributed this to the HCD methods used to develop the final system (e.g., by interviewing operators to find out what they needed from the system, then improving the interface design accordingly).

4.3.5.4   Example: Operator Visibility Assessment

Eger and colleagues (e.g., Eger et al., 2004, 2010; Godwin et al., 2008; Godwin and Eger, 2012) have demonstrated a range of techniques for assessing the visibility that operators of equipment have of their surroundings, and an understanding of how these assessment techniques may be used in virtual simulation to enhance equipment design.

4.3.5.5   Example: Manual Task Risk Assessment

Risks of musculoskeletal injury are frequently associated with the operation and maintenance of mining equipment. For example, moving and operating jackleg drills used to install secondary support in underground mines exposes the user to a high risk of musculoskeletal injury (Figure 4.9a and b). Tools are available to assess these risks and determine their acceptability (Figure 4.10).

(a) (b)

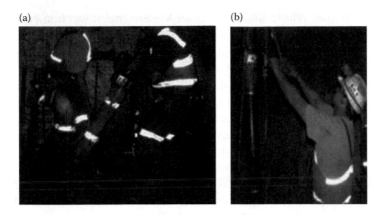

FIGURE 4.9 (a and b) Examples of manual tasks in mining.

Risk analysis: Jack-leg bolting

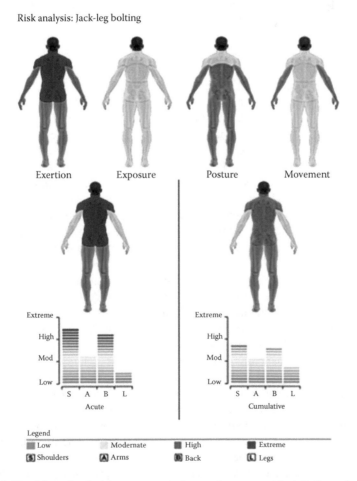

FIGURE 4.10 Manual task risk assessment of secondary support installation using a jackleg drill (ergoanalyst.com).

4.3.5.6 Example: Whole-Body Vibration Measurement (section co-written with Danellie Lynas from the University of Queensland)

Long-term exposure to high amplitude whole-body vibration is strongly associated with the subsequent development of back pain (Bernard, 1997; Bovenzi and Hulshof, 1998). Adverse consequences for cardiovascular, respiratory, digestive, reproductive, endocrine, and metabolic systems are also possible.

ISO2631-1 (ISO, 1997, 2010) describes procedures for the evaluation of whole-body vibration. Two methods of describing frequency-weighted acceleration amplitudes are commonly used to evaluate whole-body vibration exposure: (i) the root mean square (r.m.s.); and (ii) the vibration dose value (VDV). The VDV is a fourth root measure that is more sensitive to high amplitude jolts and jars. ISO2631-1 provides guidance regarding the evaluation of health effects, defining a "health guidance caution zone." For exposures below the health guidance caution zone, it is suggested that no health effects have been clearly documented. For exposures within the health guidance caution zone "caution with respect to potential health risks is indicated" and for accelerations greater than the health guidance caution zone, it is suggested that "health risks are likely." For an 8 h daily exposure, the upper and lower bounds of the health guidance caution zone are 0.47 and 0.93 m/s2 r.m.s., respectively (McPhee et al., 2009). The corresponding values for the VDV measure expressed as an 8-h equivalent [VDV(8)] are 8.5 and 17 m/s$^{1.75}$.

A range of mobile plant and equipment are used at surface and underground mines. Data collected from such equipment (e.g., Burgess-Limerick, 2012; Eger et al., 2006; Wolfgang and Burgess-Limerick, 2014a) suggests that the vibration amplitudes to which operators are exposed may lie within or above the ISO2631.1 Health Guidance Caution Zone, however the ability of mines to measure, and consequently, manage these exposures has been limited because of the expense and complexity of measurement systems. This has changed with the development and validation of an iOS application (WBV—ergonomics.uq.edu.au/wbv) that allows an iPod Touch to be employed as a simple, inexpensive method of measuring whole-body vibration amplitude (Burgess-Limerick and Lynas, 2015; Wolfgang and Burgess-Limerick, 2014b; Wolfgang et al., 2014).

Figure 4.11 illustrates 70 whole-body vibration measurements taken from dozers at a surface coal mine using the WBV application. Although about a third of the measurements exceed the ISO2631.1 health guidance caution zone, considerable variability is evident, suggesting that there may be modifiable factors contributing to the elevated vibration levels. These factors include interactions between task, geology, equipment maintenance, seat design, and operator technique (e.g., Eger et al., 2011). Work is ongoing to assess the implications of these relationships for the implementation of control measures to reduce whole-body vibration exposures. For example, in other contexts, it is clear that roadway conditions are a major contributor to elevated vibration exposures. Figure 4.12 illustrates the whole-body vibration exposures experienced by the driver of a personnel transport vehicle in an underground coal mine driving over the same section of roadway, at the same speed, before and after roadway maintenance. Measurements of this kind provide a powerful feedback loop to facilitate the human-centered design of equipment, workplaces, and tasks.

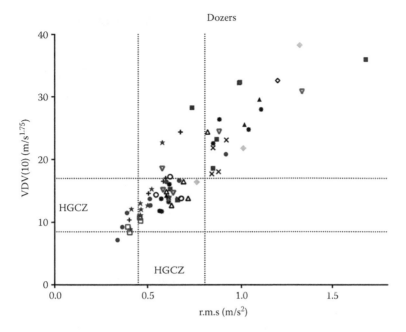

FIGURE 4.11   VDV(10) vs RMS values for each of 70 long-duration, vertical, whole-body vibration measurements taken from dozers during normal operations. Data from the same dozer on different shifts indicated by symbol. Measurement duration ranged from 140 to 660 min, median measurement duration = 440 min.

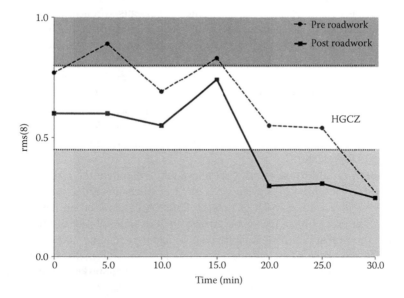

FIGURE 4.12   Vertical, whole-body vibration measurements (RMS) over five-min intervals from the driver of an underground coal personnel transport vehicle travelling over the same section of roadway, at the same speed, before and after roadway maintenance.

4.3.6   User-Error Risk Analysis in Simulated Mine Conditions

4.3.6.1   Purpose

The purpose of this step is to evaluate the possible human-error risks (operating and maintenance) of the equipment/technology in simulated mine conditions before it is actually deployed. Ideally, this would occur within the iterative design process.

4.3.6.2   Rationale and Recommended Approach

Incorporating human factors early and often in the equipment design process and life cycle is central. One way to broadly achieve this is through human-centered safe design to "design out" risks and hazards. It involves safety by design, not safety by procedure or through retrofit trial and error. Whilst this ideally would have been fully undertaken during the design stages, the opportunity to review the human-error risks of the equipment/technology in simulated mine conditions or with experienced mining personnel before it is actually deployed can be very useful. This activity has much in common with the summative evaluation that was described in Activity 5: the user-based methods described previously could also be adapted for here. In addition, two other approaches are recommended here:

1. **Prospective critical decision method**. Critical decision method (CDM) is a well-respected in-depth interview method in mining and elsewhere to investigate incidents and accidents. Horberry and Cooke (2012) extended this to also consider "prospective" incidents related to equipment use. A prospective critical decision method occurs in four stages with end users: the first stage is to brainstorm potential incidents when new equipment would be in use and select one potential incident for further analysis; in the second stage, the potential incident is decomposed into relevant decisions or actions known as "chunks" to create a timeline; in the third phase, the critical decisions for these are examined in greater depth, and the fourth and final stage involves posing a number of hypothetical scenarios/changes to the event by *what-if* questions (e.g., what if an inexperience operator was using the new device?). Prospective Critical Decision Method can help to understand what can go wrong and what potential design countermeasures might be possible.

2. **Human-reliability analysis.** One approach to estimate the error potential for different subtasks is the human error assessment and reduction technique (HEART). As noted in HCD Activity 2, HEART is based on the philosophy that every time a task is performed there is a possibility of failure and that the probability of this is affected by error producing conditions such as fatigue, distraction, and inexperience. The HEART method essentially involves four key steps: 1. Identify the full range of subtasks that the equipment operator would be required to complete for a task, 2. A nominal human unreliability score for the particular task is then determined by consulting local experts, 3. The error producing conditions that are likely to have a negative effect on the outcome are then considered and the extent to which each error producing condition applies to the task in question is discussed

and agreed upon with local experts, 4. A final estimate of the overall human-error potential is then calculated (Cloete and Horberry, 2014).

4.3.6.3   Example: Analyzing Mine Emergency Management Needs

Xiao conducted analysis of the mine emergency management needs of control-room operators in an Australian underground coal mine (Xiao et al., 2015). By applying a prospective approach similar to the prospective critical decision method, the work found that emergency management needs can be identified by capturing requirements that involved (a) emergency and routine operations and (b) social and technical aspects of the work. Xiao concluded that the approach has potential value for mining researchers and practitioners in terms of supporting the design of more effective mine emergency systems: an example of this was developing emergency scenario training for new control room operators.

4.3.7   User Manuals and Training Materials

4.3.7.1   Purpose

The purpose of this step is to prepare manuals and training materials to ensure that the equipment/technology is successfully deployed and used at a mine site.

4.3.7.2   Rationale and Recommended Approach

Regardless of how well equipment is designed, there will usually be a need to provide some degree of training and documentation for mining equipment use and maintenance. This need is particularly critical for new mining technology, which may require new operator skills and different ways of working (Lynas and Horberry, 2010). With appropriate practice and timely feedback, skills can become automatic. Training and user documentation also help provide the knowledge and cognitive skills required to undertake problem solving such as fault diagnosis, or choosing the correct response in an emergency situation.

Like many other human-centered design steps, the design of training and user manuals should encompass a structured process including front end analysis of users and tasks, an iterative design component with usability testing and evaluation. Training should include practice in a variety of circumstances and situations.

The recommended approach here involves three interlinked activities (Tichon, 2011):

1. **Front-end analysis (or training needs analysis)**: A full analysis of the tasks to be performed by operators and maintainers, the use context, the equipment interface, the trainees, and their training needs and resources, leading to a definition of the training functional specifications.
2. **Design and development**: Training concept generation, iterative training system development and prototyping, and usability testing. The most effective instructional strategies embody four basic principles: (a) the presentation of the concepts to be learned; (b) demonstration of the knowledge, skills and behaviors required; (c) opportunities to practice; and (d) feedback during and after practice.

3. **Training system evaluation**: Determining training evaluation criteria (e.g., performance time), collection and analysis of these data, and subsequent modification of the training system, if indicated.

4.3.7.3 Automotive Example: User-Centered Design of Assembly-Line Training

This work used a human-centered-design–style approach to develop a training system for vehicle assembly line operators (Hermawati and Lawson, 2013). The impetus for this was the extra pressure and operator skills requirements that exist today due to the increasingly competitive nature of the automotive industry and the current demand for assemble-to-order products. The researchers undertook four interrelated activities here:

1. **Establish the context of use.** They collected data from end-user interviews, observations, and personnel records to identify the main user groups (e.g., trainee, engineer, and trainer/foreman).
2. **Specification of user requirements.** Additional interviews, focus groups, and field observations were conducted leading to task and stakeholder analyses to inform the development of the training systems.
3. **Producing design solutions.** Hermawati and Lawson (2013) used scenarios to depict how the prototype training system would be used in different circumstances for the user groups.
4. **Evaluation.** The researchers iterated the training system design through extensive end-user feedback.

This approach resulted in an effective and acceptable training system being developed for vehicle assembly line operators. So, the key message here is that focusing on users and their tasks and iteratively designing training systems was found to be an effective process.

4.3.7.4 Example: Competency Assessment

Human-centered design can also be applied to assessing the outcomes of training. A surface coal mine recently made a decision to change the road rules in use at the site to introduce a set of "priority rules" to govern give-way procedures. Under these rules, light vehicles were required to always give way to heavy vehicles such as haul trucks. The change has potential to increase both productivity (loaded haul-trucks being required to stop less frequently) and safety (light vehicles having better visibility of haul trucks than vice-versa). Prior to making the change, the site put a training process in place for both light- and heavy-vehicle drivers. In collaboration with university researchers, an online video-based competency assessment test (analogous to video-based hazard perception tests, e.g., Horswill et al., 2015) was constructed in which trainees were presented with real-world video clips illustrating a range of scenarios and were required to indicate whether a "give-way" or "proceed" response was appropriate. The test was used to provide a principled way of determining whether trainees were permitted to drive onsite or whether further training was required.

4.3.8   Integration of the Equipment with Other Work Systems

4.3.8.1   Purpose

The purpose of this step is to effectively integrate the equipment/technology with existing work systems at a mine site; that is, with other equipment, tasks, and the organizational and physical environment. This builds on knowledge gained about the context of use identified in Activity 1.

4.3.8.2   Rationale and Recommended Approach

New equipment or technology being deployed at a mine site often requires different methods of working (Horberry, 2012). This might be as simple as different access procedures on a new design of a bulldozer or a completely different way of working for new, automated mining systems. For all, operator acceptance is a key issue, and poor operator acceptance of equipment/technology after deployment might lead to mistrust, incorrect use, or even sabotage.

Human-centered deployment can help reduce problems with new mining technology acceptance. It can be thought of as part of human systems integration (HSI): an overall approach for integrating the domains of human factors engineering, system safety, training, personnel, crewing, health hazards, and survivability into each stage of the systems life cycle (Burgess-Limerick et al., 2011; Stanton et al., 2013). This wider topic related to mining equipment and new technology is further considered in the final chapter of this book.

However, leaving the determination of how (and whether) the new mining equipment can be integrated with existing equipment and work systems until after it has been designed and manufactured is way too late. The process needs to begin with understanding the likely use context (presented in Activity 1). One approach here involves four interlinked activities (Horberry and Cooke, 2014):

1. **Understanding tasks and the use context.** As with many other HCD steps, a full task analysis for equipment operation and maintenance is often vital before the equipment/technology is deployed.
2. **Understanding end users.** Understanding the physical and cognitive capabilities of the operators and maintainers. A key issue here is ensuring that they have sufficient skills and training to effectively use and maintain the equipment/technology.
3. **Operator consultation.** Consultation can include discussing the need for the new equipment/technology and addressing operator concerns prior to deployment. After deployment, regular operator consultation/ feedback is vital to make system improvements and to ensure continuing acceptance.
4. **Management commitment and feedback.** Management commitment (from supervisors to mine managers) is vital throughout the deployment process. Support for the new technology might include changing procedures, seeking specific feedback, and giving explicit training.

4.3.8.3   Example: Operator Acceptance of Seatbelt Interlocks for Forklift Trucks

Work by Horberry et al. (2004, 2006) examined forklift seatbelt "interlocks" that prevent a vehicle from starting until the seatbelt was correctly engaged. Ensuring that the operators accepted the technology was vital. Horberry et al knew that effective solutions cannot be developed without first understanding the issues: these included the likely users, the forklift tasks, and the use environment. From their task analysis, user requirements capture and consultations with workers and managers, they anticipated that the interlocks would be successful only if they met the following five points:

1. Be technologically feasible and not require overly extensive retrofitting
2. Be supported by managers and not be strongly opposed by operators
3. Be capable of integrating with other equipment, training regimes, and safety procedures
4. Have no "low-technology" countermeasure equally capable of performing the same role
5. Be reliable, display the required information, and produce few false alarms

After deployment, they found that these five points were valid: the seatbelt interlocks worked well and were generally accepted by the forklift drivers. Understanding users and tasks, and regular consultation with workers and managers were vital issues for successful seatbelt interlock system integration.

4.3.8.4   Example: Impact on Operators of Adding Proximity Detection to Continuous-Mining Machines

Hass and Rost (2015) interviewed nine operators of continuous-mining machines fitted with proximity detection systems to examine the consequences of the introduction of the technology. The operators reported standing in "red" danger zones less frequently; however, they also reported that considerable changes to their operating strategies were required to adapt to the new technology. The operators suggested that the introduction of the technology required additional training support to gain a better understanding of the functionality of the system and the ways it integrated with the operation of the continuous-mining machine.

4.3.9   IDENTIFY UNWANTED EVENTS IN ACTUAL OPERATIONAL CONDITIONS

4.3.9.1   Purpose

The purpose of this step is to identify, investigate, and propose countermeasures to design-induced user error events that occur in actual operational conditions.

4.3.9.2   Rationale and Recommended Approach

The dominant view in safety science today is that "human error" is best viewed as a consequence, not as a cause (Simpson et al., 2009). So, related to mining equipment and technology, user error should be seen as a consequence of design deficiencies.

Identifying, investigating, and proposing countermeasures to design-induced user-error events that occur in actual operational conditions is consequently a vital component of mining HCD. Two approaches are recommended here: in addition, simpler methods such as log books for operators to quickly note down any perceived design deficiencies whilst they are on shift can also be of considerable assistance.

1. **Safety in design ergonomics (SiDE).** The safety in design ergonomics process was described in Activity 3 of the overall HCD process for participatory design with end users, but it can also be used to explore design-induced user-error events that occur in operational conditions. It usually involves seven stages. Stages 1–4 are usually performed in joint designer/end-user workshops. Stage 0 defines the context and scope of the whole process. Stage 1 identifies the critical tasks; for mining equipment, a list of maintenance and operational tasks performed may already exist. From the task lists, critical tasks are prioritized. Stage 2 involves decomposition of the high priority tasks identified in Stage 1. Stage 3 involves identification of risks (and user errors) for each of the subtasks in Stage 2. Stage 4 is the development of human-centered solutions for the risks/errors identified in Stage 3. Stage 5 involves iterative development of the Step 4 solutions. Stage 6 involves managing and documenting.
2. **Critical decision method (CDM).** CDM is an emerging in-depth method in mining to investigate incidents. It involves interviewing the active decision makers in the event. It occurs in four stages: the first stage is to identify incidents when the equipment was in use and select one incident for further analysis; in the second stage, the incident is decomposed into relevant decisions or actions known as "chunks" to create a timeline; in the third phase, the critical decisions are examined in greater depth using predefined "probes"; and the fourth and final stage involves posing a number of hypothetical scenarios/changes to the event (e.g., what if an inexperience operator was involved?). From this, redesigns can be identified using a process called decision centered design: this uses the key decision points as a basis for interface design and skills training to best assist the operator.

4.3.9.3 Example: Improving Mobile Mining Equipment Access

A high number of access injuries can occur with mobile mining equipment (Horberry et al., 2016). This work used a SiDE-type process in a workshop format with participants from global advisory groups, site management, and operators and maintainers.

The workshop began by identifying the types of access tasks, and the different environments in which they are performed at the mines. One high priority task was accessing the truck cab.

This task was viewed by the workshop participants at the mine site: an experienced operator and an inexperienced person accessing the equipment were observed at the mine site. In the workshop, the task was decomposed, risks were analyzed, and solutions to minimize user error when accessing the truck cab were proposed. One of these solutions involved the provision of additional hand guiderails for use during access and egress from the equipment. This was iterated and improved through user

FIGURE 4.13   Improved design of access guiderails.

trials (e.g., to always have three points of contact, Figure 4.13), and the design was then used at the mine site for this equipment type.

4.3.10   FEEDBACK TO DESIGNERS: FOR THE NEXT GENERATION OF THE EQUIPMENT

4.3.10.1   Purpose

The purpose of this step is to feedback equipment design deficiencies to designers to improve future generations of the equipment.

4.3.10.2   Rationale and the Recommended Approach

Designers of mining equipment are sometimes unable to see their equipment in actual use at mines (Cooke, 2015). Design may even take place on the other side of the world from the mine site. Feedback to designers and manufacturers about the features that were unsuccessful (and successful) is consequently vital for improvements to future equipment generations. Some manufacturers have customer feedback processes; however, often this is only text-based and must pass through many hands before it reaches the appropriate designer (Horberry and Cooke, 2012). By this stage, it may be difficult to understand the precise issue, and the person who raised it may not be contactable. So three approaches are recommended here:

1. **Visits to mine sites.** The most direct and perhaps most effective approach is for designers to visit mine sites and talk to workers. Ideally this would occur regularly. However, the sheer number of sites, their variety, and their remoteness may make the process difficult.
2. **Direct customer feedback of current products.** Establishing a formal method for operators and maintainers to directly report design deficiencies

back to the manufacturers is vital. It needs to be detailed enough to understand the precise design issue. Ideally it would also contain contact details of the mine site person reporting it, so the designer can follow up.

3. **Video of tasks**. As seen in the following example, obtaining end-user inputs by means of recording current tasks on video can be vital for seeing how equipment is actually used at mines.

4.3.10.3 Example: Feedback to Designers at a Major Mining Equipment Manufacturer

Horberry and Cooke (2012) undertook interviews with designers from a major manufacturer of mining equipment. They asked which formal and informal human-related methods they used to create and assess the safety of their equipment, and how they obtained end-user feedback.

They found that the designers used a range of human-centered methods to improve the safety of their equipment, including an ergonomic checklist for the operators' cabin, in-field usability observation, customer feedback and risk assessment workshops. However, access to end users in the workshops and the limited number of in-field observations sometimes restricted the designers from knowing what really happens in mine site conditions.

Obtaining end-user inputs by video recording current tasks was viewed by designers as extremely beneficial. Video records can provide objective visual information that is easily shared, shows the designers how their equipment is actually used, acts as a memory aid for those designers who have been to a mine site, and, more generally, can be an excellent reminder that real people will eventually use their equipment.

Video records are, of course, not an adequate replacement entirely for direct end-user feedback (e.g., observations and interviews): the designers would be unable to ask the person about the task and so may have to interpret the actions on the video. Equally, other design options could not be explored by this method, and if a design is changed significantly then the task itself often changes. Despite these caveats, the use of videos of tasks was judged to be a key human-centered method that could be used in addition to the four other ergonomics approaches designers currently use.

4.3.11 HCD Summary Report

4.3.11.1 Purpose

The purpose of this step is to help manufacturers and designers develop a summary report of human-centered issues for potential mine site purchasers of equipment/technology to consider.

4.3.11.2 Rationale and the Recommended Approach

The use of an HCD summary report by mining equipment manufacturers to show how they have addressed human factors issues can be vital. It is similar to the USA FDA process for human-centered medical device designs and to some mining equipment regulations in Australia (e.g., as required in MDG-15). Human-centered design summary reports are now also being used by mining companies: requiring their equipment suppliers to provide the report during procurement (see the following "EDEEP" example).

The recommended approach is for mining equipment developers to use a structured and consistent template such as the one shown in Table 4.1.

4.3.11.3 Example: EDEEP (EMESRT Design Evaluation for Equipment Procurement)

EDEEP is a process to allow OEMs to provide information to the purchasers of earth moving equipment: demonstrating their actions towards minimizing

---

TABLE 4.1

Recommended Mining HCD Summary Report Template

1. Intended device users, uses, use environments, and training	• Intended user populations and differences in capabilities for multiple user groups • Intended uses, use environment and operational contexts of use • Training intended for users
2. Device user interface (for new mining technologies)	• Graphical depiction (drawing or photograph) of device user interface • Verbal description of device user interface
3. Summary of known use problems	• Known problems with previous (or similar) models • Design modifications implemented in response to user difficulties
4. User task selection, characterization and prioritization	• Risk analysis methods • Use-related hazardous situation and risk summary • Critical tasks identified and included in user validation tests
5. Summary of formative evaluations	• Evaluation methods • Key results and design modifications implemented
6. Validation testing	• Rationale for test type selected (i.e., simulated use or evaluation), test goals • Number and type of participants and for how they represent the user populations • Test results: Device uses, unanticipated use errors, success and failure occurrences • Subjective assessment by test participants of critical task difficulties • Description of task failures, implications for additional risk mitigation
7. Conclusion (The preceding methods and results support this conclusion)	The <Name Model> has been found to be reasonably safe and effective for the intended users, uses and use environments. Any residual risk that remains after the validation testing would not be further reduced by modifications of design of the user interface, is not needed, and is outweighed by the benefits that may be derived from the equipment's use.

Source: FDA (2016a).

and mitigating risk within maintenance and operability tasks through the use of design controls. The EDEEP document is made up of four key sections to be supplied to the purchaser for evaluation (and is thus similar to an HCD summary report).

1. Critical task identification information
2. Design philosophy reference information (e.g., visibility)
3. Task-based risk assessment information
4. Design feature information from the task-based risk assessment

The global equipment manufacturer Sandvik used a version of EDEEP to review the design of one piece of equipment and to evaluate how potential safety risks could be mitigated. Areas that were successfully addressed in this process included operator visibility, ease of maintenance and operation, and ease of equipment access. Sandvik viewed the EDEEP process as being very beneficial.

4.4 SUMMARY OF HCD ACTIVITIES

The 11 human-centered design activities described earlier in this chapter, their purposes, and the methods used to achieve them are summarized in Table 4.2.

TABLE 4.2
Summary of the 11 HCD Activities for Mining Equipment/Technologies

Activity	When to Be Undertaken	Purpose	Recommended Methods
1. Understanding the need for the equipment, the likely users and the use environment	Conceptual design and early in the design process	To understand the need for the equipment from an end-user perspective, who will be using the equipment, what are their requirements and what is the likely use context	• Observe users to determine what they do, and what they really need • Understand the likely users and specify user requirements • Describe how and where the equipment would be used
2. Task, incident and error analysis for the equipment	Early in the design process	To understand the likely tasks with equipment/technology, and what has gone wrong (errors or incidents) with similar systems	• Task analyses of equipment operation or maintenance • Incident and error analysis
3. Participatory design with end-users	Early in the design process	To work with end users to design new versions of equipment/technology	• Decision centered design • Safety in design ergonomics/OMAT

(Continued)

TABLE 4.2 (*Continued*)

Summary of the 11 HCD Activities for Mining Equipment/Technologies

Activity	When to Be Undertaken	Purpose	Recommended Methods
4. Application of relevant standards (e.g., ISO, ANSI) and human factors data	Through the design process	To apply relevant standards, guidelines and human factors data to the design of the equipment/ technology	• Anthropometric data set/ models • Manual tasks • Work-station design • Control design • Display design principles
5. Testing and evaluation of equipment/ technology	Through the design process	Both to test the design as it progresses, and to evaluate the completed equipment from a human-centered perspective	• Formative evaluation • Summative evaluation: inspections based • Summative evaluation: user based
6. User error risk analysis in simulated mine emergency conditions	At mine site deployment	To evaluate the possible human error risks of the equipment in simulated mine conditions before it is actually deployed	• Prospective critical decision method • Human reliability analysis
7. User manuals and training materials	At mine site deployment	To help prepare manuals and training materials to help ensure that the equipment is successfully deployed at a mine site	• Front-end analysis (or training needs analysis) • Design and development • Training system evaluation
8. Integration of the equipment with other work systems	Before, at and after, mine site deployment	To effectively integrate the equipment with existing work systems at a mine site	• Understanding tasks and use context • Understanding end users • Operator consultation • Management commitment and feedback
9. Identify unwanted events in actual operational conditions	During routine operation	To identify, investigate and propose countermeasures to design-induced user error events in operational conditions	• Safety in design ergonomics • Critical decision method
10. Feedback to designers for next generation of the equipment	During routine operation/ modification	To feedback deficiencies to designers to help improve future generations of the equipment	• Visits to mine sites • Direct customer feedback of products • Videos of tasks
11. HCD summary report	At procurement	To develop a summary report of human-centered issues for potential mine site purchasers of equipment to consider	• HCD summary report template for mining technology/equipment manufacturers to show how they have addressed human factors issues

4.4.1 HCD Scenarios

We recognize that there is no single "standard" equipment or technology design process in mining. Therefore, two examples are given here for how the preceding human-centered design material could be used.

4.4.1.1 To Develop Mining Technologies and Equipment by R&D Professionals

For mining equipment and technology developers, the areas ticked in Table 4.3 would be particularly relevant.

Only one activity is less relevant here: Step 9 would fully occur at the mine site itself. Even then, it could be argued that this mine site operational information would be vital to help iterate future generations of the equipment/technology.

TABLE 4.3

The 11 HCD Activities for Developers of Mining Technologies and Equipment

	Activity	When to Be Undertaken	Purpose	Recommended Methods
✓	1. Understanding the need for the equipment, the likely users and the use environment	Conceptual design and early in the design process	To understand the need for the equipment from an end-user perspective, who will be using the equipment, what are their requirements and what is the likely use context	• Observe users to determine what they do, and what they really need • Understand the likely users and specify user requirements • Describe how and where the equipment would be used
✓	2. Task, incident and error analysis for the equipment	Early in the design process	To understand the likely tasks with equipment/technology, and what has gone wrong (errors or incidents) with similar systems	• Task analyses of equipment operation or maintenance • Incident and error analysis
✓	3. Participatory design with end-users	Early in the design process	To work with end users to design new versions of equipment/technology	• Decision-centered design • Safety in design ergonomics/OMAT
✓	4. Application of relevant standards (e.g., ISO, ANSI) and Human factors data	Through the design process	To apply relevant standards, guidelines and human factors data to the design of the equipment/technology	• Anthropometric data set/models • Manual tasks • Work-station design • Control design • Display design principles

(Continued)

TABLE 4.3 (*Continued*)

The 11 HCD Activities for Developers of Mining Technologies and Equipment

	Activity	When to Be Undertaken	Purpose	Recommended Methods
✓	5. Testing and evaluation of equipment/ technology	Through the design process	Both to test the design as it progresses, and to evaluate the completed equipment from a human-centered perspective	• Formative evaluation • Summative evaluation: inspections based • Summative evaluation: user based
✓	6. User error risk analysis in simulated mine emergency conditions	At mine site deployment	To evaluate the possible human error risks of the equipment in simulated mine conditions before it is actually deployed	• Prospective critical decision method • Human reliability analysis
✓	7. User manuals and training materials	At mine site deployment	To help prepare manuals and training materials to help ensure that the equipment is successfully deployed at a mine site	• Front-end analysis (or training needs analysis) • Design and development • Training system evaluation
✓	8. Integration of the equipment with other work systems	Before, at and after, mine site deployment	To effectively integrate the equipment with existing work systems at a mine site	• Understanding tasks and use context • Understanding end users • Operator consultation • Management commitment and feedback
X	9. Identify unwanted events in actual operational conditions			
✓	10. Feedback to designers for next generation of the equipment	During routine operation/ modification	To feedback deficiencies to designers to help improve future generations of the equipment	• Visits to mine sites • Direct customer feedback of products • Videos of tasks
✓	11. HCD summary report	At procurement	To develop a summary report of human-centered issues for potential mine site purchasers of equipment to consider	• HCD summary report template for mining technology/equipment manufacturers to show how they have addressed human factors issues

4.4.1.2 For Regulators, Mine Site Managers, and Procurement Managers

For these groups, there may be less of a need to go into excessive detail for many of the earlier HCD steps. Depending on the exact context, the areas ticked in Table 4.4 may be particularly relevant.

TABLE 4.4

The 11 HCD Activities for Regulators, Mine Site Managers and Procurement Managers

	Activity	When to Be Undertaken	Purpose	Recommended Methods
X	1. Understanding the need for the equipment, the likely users and the use environment			
X	2. Task, incident and error analysis for the equipment			
X	3. Participatory design with end-users			
X	4. Application of relevant standards (e.g., ISO, ANSI) and human factors data			
X	5. Testing and evaluation of equipment/technology			
✓	6. User error risk analysis in simulated mine emergency conditions	At mine site deployment	To evaluate the possible human error risks of the equipment in simulated mine conditions before it is actually deployed	• Prospective critical decision method • Human reliability analysis
✓	7. User manuals and training materials	At mine site deployment	To help prepare manuals and training materials to help ensure that the equipment is successfully deployed at a mine site	• Front-end analysis (or training needs analysis) • Design and development • Training system evaluation
✓	8. Integration of the equipment with other work systems	Before, at and after, mine site deployment	To effectively integrate the equipment with existing work systems at a mine site	• Understanding tasks and use context • Understanding end-users • Operator consultation • Management commitment and feedback
✓	9. Identify unwanted events in actual operational conditions	During routine operation	To identify, investigate and propose countermeasures to design-induced user error events in operational conditions	• Safety in design ergonomics • Critical decision method
✓	10. Feedback to designers for next generation of the equipment	During routine operation/modification	To feedback deficiencies to designers to help improve future generations of the equipment	• Visits to mine sites • Direct customer feedback of products • Videos of tasks
✓	11. HCD summary report	At procurement	To develop a summary report of human-centered issues for potential mine site purchasers of equipment to consider	• HCD summary report template for mining technology/equipment manufacturers to show how they have addressed human factors issues

In this situation, the latter HCD steps are most important, after the initial design is largely complete. Preparing for deployment, refining user manuals/training, integration with other work systems, identifying operational errors, giving feedback to designers, and the HCD summary report are of importance here. For procurement managers, the final human-centered design summary report should be a key document.

To further build on the preceding information, we provide detailed mining HCD case studies linked to these eleven activities in the next chapter.

5 HCD Case Studies

5.1 OVERVIEW

This chapter presents detailed HCD case studies from the global minerals industry. Each case study follows a similar format: introduction, case study description, and then human-centered design analysis using the 11 possible HCD activities. Taken as a whole, these case studies show how HCD-style approaches have been used successfully for mining equipment and new technology in the past. Regarding the greater future use of human-centered design, all sectors of the minerals industry—OEMs, technology developers, mine sites, and regulators—can help achieve safety and productivity gains through the employment of HCD-style approaches.

5.2 CASE STUDY 1: HCD WITH UNDERGROUND LOADER AUTOMATION AT CMOC NORTHPARKES, AUSTRALIA

5.2.1 INTRODUCTION

Automation offers the minerals industry great potential for improvements in productivity and safety. However, the experience of introducing automation in other industries has been that the full potential of new technology is not always realized (see e.g., Lee and Seppelt, 2009). Designers are surprised, for example, to find that automation does not eliminate human errors. Unwanted and unexpected consequences also arise if the introduction of automation fails to consider how people will adapt to the new technology. A focus on the technical aspects of automation is necessary but not sufficient for success. The potential for improvements in productivity and safety promised by automation will only be achieved if the joint cognitive system that emerges from the combination of humans and automation is designed to perform the functions required for system success (Woods and Hollnagel, 2005).

Automation can be defined as a technology that performs a function previously carried out by a person (Parasuraman and Riley, 1997). It has the potential to allow the joint human-automation system to achieve levels of performance and safety that are otherwise impossible. For example, the introduction of mechanical cutting of coal to replace manual labor was an early form of mining automation.

Examples of automation introduced to mining more recently include

- Vehicle proximity warning systems installed on surface haul-trucks
- Fatigue detection systems, e.g., smartcap
- Software for mine planning and enterprise optimization
- Pedestrian proximity detection systems interlocked with underground continuous mining machines

- Automatic face alignment and horizon control of underground coal longwall equipment
- Automatic cutting cycles of continuous mining machines
- Automation of the swing, dump, and return phases of the shovel loading cycle, as seen earlier in the work of Cloete and Horberry (2014)
- Automated drilling systems and automated haul-trucks at surface mines
- Automated haulage in underground metal mines, an example of which is the subject of this case study

5.2.1.1   Changing Roles of People in the System

The addition of automation does not eliminate people from the system. Rather, the roles of people are changed and new tasks are introduced. The importance of these new tasks are frequently underestimated, particularly the need for people in the system to respond to unanticipated situations, which may include malfunction or failure of an automated component of the system.

Automation sometimes changes the role people play in the system from continuous active control of a system component to passive supervision or monitoring. One consequence of this change can be degradation of the manual control skills of operators. Introducing automation also often changes the type and extent of information available to equipment or plant operators by removing them from direct contact with the process being controlled. This reduces the sources of information (visual, auditory, haptic, olfactory) that may be used to monitor the system and, in particular, to detect and diagnose the causes of departures from normal operation. Both the change from manual control and the reduced information directly available to the operator, potentially leads to loss of situation awareness, and response delays in the event that a human operator is required to take action in response to the system being perturbed beyond its normal operating range.

5.2.1.2   Maintaining Situation Awareness

A challenge for the designer is to ensure the operator maintains situation awareness by determining what information is required by the operator and how this may be provided without overwhelming the operator with data. The design of the interfaces by which information is conveyed becomes a critical concern. Combining data into meaningful information though the creation of visual displays with emergent properties that correspond to system relevant parameters is one approach that may be helpful, as is placing information in a meaningful context (indicate allowable ranges, for example), and/or integrating automation-related information with traditional displays (see e.g., Flach et al., 1998). Displays should avoid the need for the operator to undertake mental transformations to gain meaning from the information. Other options are to create interfaces that predict future states of the system, and/or provide information through multiple sensory channels.

5.2.1.3   Avoiding "Clumsy Automation"

A potential trap associated with the design of automation is so called "clumsy automation" (Wiener, 1989) in which easy tasks are automated while complex tasks are left for a human operator, sometimes because they are too difficult to automate.

This can exacerbate the loss-of-situation-awareness issue noted earlier. It can also mean that workload is reduced during already low-workload phases of work, while remaining unchanged, or even increased, during high-workload operations because of the cognitive overhead associated with engaging and disengaging automation (Kirlik, 1993).

The choice of which of the functions currently performed by humans should be automated is a nontrivial question that deserves careful consideration. People are good at perceiving patterns. They adapt, improvise, and accommodate quickly to unexpected variability. People are not good at precise repetition of actions, or vigilance tasks. However, designing the system requires more than allocating functions to person and machine—rather, the challenge is to identify how the operator and automation can jointly perform the functions required for system success.

5.2.1.4   New Types of Errors

The potential for new types of operational errors, such as configuration errors or errors arising from to mode confusion, may be introduced by adding automated components to a system. At the same time, the span of control of an individual operator is likely to be increased by the automation. Timely feedback to the operators about the system's status is vital: delays in receiving feedback can increase the potential for errors. If a reduction in crewing occurs as a consequence of the introduction of automation then there is also reduced redundancy, which may further reduce the probability of error detection and correction. Catastrophic outcomes can therefore result from this combination of automation consequences.

5.2.1.5   Human Responses to Automation Can Have Unanticipated Consequences

The response of humans to the introduction of automation can also lead to unanticipated consequences. One dimension of the human responses relates to the trust the human operator has in the automation technology. Operators may come to be complacent and overtrust the automation, either failing to note and respond to automation failures (particularly when such failures are rare) or altering behavior in ways that reduce the intended safety benefits of automation. For example, the proposed introduction of pedestrian proximity detection technology interlocked with the braking systems of underground coal haulage equipment could potentially lead to operators and pedestrians taking less care to avoid interactions, with potentially fatal consequences.

Conversely, lack of trust in automation may lead to operators disengaging, disabling, or ignoring the technology. For example, a high rate of false-alarms is a threat to the introduction of any proximity detection technology, and in general, the failure of technology to improve short term productivity is a threat to the adoption of the technology.

The change in operator roles associated with the introduction of automation can also be problematic if the operator's job satisfaction is reduced. The key here is to allow operators to leverage old skills into new ones, as well as involving operators in the automation design and implementation process and empowering them to have an ongoing role in improving the system.

5.2.2   LOADER AUTOMATION AT CMOC NORTHPARKES

The aim of this case study is to investigate the successful implementation of loader automation at CMOC Northparkes through presenting information gained through observation and interviews with operators and CMOC Northparkes staff.

Northparkes is a copper/gold block caving operation in central New South Wales, Australia, which commenced mining in 1994. The mine is owned by CMOC (80%) and Sumitomo Group. Teleoperation, and subsequently automation, of the underground loaders has been trialed in various forms at the mine since 1998. Implementation of the Sandvik Automine system currently in use was initiated in 2010.

Operators located in a surface control room (Figure 5.1) load ore at draw point using manual teleoperated control. The loader is then switched to automated mode to travel to the Run-of-mine bin where the ore is dumped autonomously. The loader then returns to the next draw point selected autonomously.

Each of three operators is typically responsible for three loaders. The operators' interface with the system is via three screens, keyboard and mouse, as well as joysticks and pedals that mimic the controls found in a manually driven loader (Figure 5.2). Audio from the loader is available to the operators, and the operators communicate with other crew members underground via radio. One screen (Autonomous Control System) provides the operators with information and control over the overall system, whereas a second (Mission Control System) allows the operator to monitor the location of loaders, select loaders for manual control, and modify the drawpoints to which a loader will travel autonomously. A third screen provides a video feed from the loader (switchable between front and rear) and a schematic "teleoperation assist" window that provides an indication of the location of the loader being controlled or monitored relative to the laser scanned surroundings (Figure 5.3).

FIGURE 5.1   Surface control room layout. Two autonomous loader operators are located at the front of the room.

FIGURE 5.2 Autonomous loader workstation.

FIGURE 5.3 Video feed and "teleoperation assist" window provided to operators.

As well as undertaking manual teleoperated loading at the draw point, the operators are also responsible for monitoring the overall system status and making decisions in response to events such as loader breakdowns, typically modifying the planned sequence of draw points. The operator can also modify the behavior of the loader during autonomous phases. For example, if a section of roadway has deteriorated, an operator can highlight the roadway section via the Mission Control Interface and set a reduced maximum speed for the loader during the autonomous travel in that section of roadway.

Implementation of the autonomous loaders has reduced the exposure of operators to a range of injury risks, particularly whole-body vibration and musculoskeletal injury risks. Productivity benefits include the ability to continue to mine through shift changes and blasting, resulting in a 23% improvement in daily tones produced.

Information obtained through a site visit and interviews with operators, safety staff, and the project manager revealed the following strategies for successful design and implementation of the automation system.

5.2.2.1   Involve the People Who Will Be Impacted

Involving the people who will be impacted by the change was seen as critical. An initial step in the implementation of the current automation system was for representatives of all stakeholders to spend three days mapping out the consequences of the proposed automation for all tasks undertaken across the mine. It became clear during this process that all underground tasks would be affected. For example, at shift change the impact of the introduction of surface control of loaders enables production to continue. This removes time pressure to execute the change, allowing greater time to be spent in shift handover and planning for the new shift. However, it was noted that access to, or through, sections of the mine where autonomous loading was in operation would be prevented, which impacted the performance of a range of other tasks.

5.2.2.2   Constant Communication between Operators and Designers

Constant communication between operators and designers throughout the implementation and subsequent operation of the autonomous system has been critical in developing and refining the user interface. The continuous presence of OEM expertise on site allows a rapid feedback loop with designers.

5.2.2.3   Provide Operators with the Information They Need

Providing operators with opportunities to suggest modifications to the system has been a key feature in the success of the automation implementation. Operators continually update a list of issues, and a "wish list" of improvements, which are fed back to the system designers, resulting in many changes to the system. For example, equipment damage was occurring because the loader was hitting the walls of the draw point while under manual control. A suggestion made was to use the laser scanners already in place for autonomous navigation to detect the proximity of the walls during manual operation, and to convey this information to the operators through changes in color of the scanning information provided on the teleoperation assist window. This information was also used to automatically apply the loader's brakes prior to the loader hitting the walls.

Similarly, wheel spin causes damage to the loader's wheels but was hard for operators to detect while loading remotely. A wheel-slip detection sensor was added and an indication of wheel slip provided to the operator through a change in color of the schematic loader wheels in the teleoperation assist window. In both cases the presentation of relevant information to the operators in a meaningful way ensured the information could be used effectively to reduce equipment damage.

Relevant information is also conveyed inadvertently, rather than by design. One operator explained that it can be difficult to gauge when the bucket has been lowered sufficiently to the ground in preparation for loading. If too much pressure is placed on the ground by the bucket, the loader's front wheels will raise and wheel slip occurs. The operator noted that the camera shake, which could be seen on the video feed when the bucket was lowered, was a useful cue.

5.2.2.4   Avoid Providing Information They Do Not Need

Conversely, another change made was to reduce the number of loader fault alarms presented to the operator. Many of these alarms, while relevant to an engineer during commissioning, are not relevant to the day to day operation of the loader. As well as being a nuisance to operators because each message required acknowledgement, being habituated to frequent non-essential error messages led on at least one occasion to an operator failing to react to a critical error, with potentially serious consequences. Presenting only essential information is another human-centered design principle.

5.2.2.5   Provide the Operators with Flexibility

Providing flexibility in information provision is another key strategy. In this case, the loaders are fitted with a microphone, and the audio is available to the operators; however, this information is not wanted by the operators, and the audio is not switched on because the nuisance value of the noise outweighs the benefit of any relevant information conveyed. Opportunities for improving the provision of information to the operators are continually being explored; for example, the camera system will soon be updated to high definition.

Many details of the automation implementation were left to production crews to determine. For example, in the transition to autonomous loading, some crews decided to ensure all crew members were trained for autonomous control, while others chose to have specialist autonomous operators. The number of loaders for which an operator should have responsibility was also determined by the crews. Whilst four loaders can be controlled by one person, the cognitive load was overly fatiguing and three was determined to be optimal. During operation, some crews choose to allocate three loaders to be controlled by each operator, while other crews allowed more flexibility, with all loaders able to be controlled by any of the three operators on shift at any one time.

Allowing crews to choose different strategies provides opportunity to evaluate different options, and comparisons between operator and crew productivity can be used to fine-tune operator strategies and identify aspects of operator behavior that lead to improved productivity.

Interesting questions remain about the training of future operators, such as whether autonomous loader operators need to necessarily have prior experience of manual loader operation, and if not, how new autonomous loader operators will be trained. The development of an autonomous loader simulator may be a useful technology to use to select and train future operators, who may not necessarily have an underground mining background.

5.2.2.6   Empower Operators to Take Action

In some cases production crews also have taken action without involving the system designers. One issue being encountered was that the cameras and scanners were accumulating dust, which was causing the automation to fail. While the system designers were exploring options for on-board cleaning mechanisms, the crews devised a means of dumping water on the camera and scanners when required. In another case, the mine's engineering team implemented emergency stop sensor under the loader independently of the manufacturer as an additional safety feature.

Making all aspects of the control system and flexible as possible and giving operators maximum control over the automation increases the degrees of freedom operators have to adapt to new situations. This potentially conflicts with design engineers' natural tendency to constrain the system and restrict operator discretion.

5.2.2.7   Take Advantage of New Possibilities

The implementation of autonomous loading has also had unanticipated consequences for future process improvements. The ability to more flexibly execute different draw point extraction patterns and modify these extraction patterns, has prompted the development of optimization software to determine in real-time the optimal pattern of extraction. This is itself a form of automation that will provide assistance to the shift-boss in maintaining situation awareness of the extraction and aid in decision making.

5.2.3   HCD Analysis

Table 5.1 compares the investigation of the successful implementation of loader automation at Northparkes with the 11 human-centered design activities introduced in the previous chapter of this book.

TABLE 5.1
Comparison of Northparkes Case Study Observations with the 11 Possible HCD Activities

Activity	Application at Northparkes
1. Understanding the need for the equipment, the likely users and the use environment.	*Undertaken.* The current automation was developed iteratively over a decade. It gained an understanding of the need for the equipment from an end-user perspective by means of representatives of all stakeholders spending three days mapping out the consequences of the proposed automation for all tasks undertaken across the mine. The system was tailored to the mine. The system engineer and Sandvik engineering gained an understanding of who would be using the system, what their requirements are, and what is the likely context of use through regular observations and discussions with operators.
2. Task, incident and error analysis for the equipment	*Undertaken.* As noted in step 1, all the underground tasks were analyzed as part of the three-day stakeholder workshop.
3. Participatory design with end users	*Partially undertaken.* End users were involved in the redesign of aspects of the system. Although much of the original design by Sandvik was conducted in Finland.
4. Application of relevant standards (e.g., ISO, ANSI) and HF data	*Undertaken.* The original design by Sandvik was conducted in Finland taking into consideration relevant standards, guidelines, and relevant human factors data.

(Continued)

TABLE 5.1 (*Continued*)

Comparison of Northparkes Case Study Observations with the 11 Possible HCD Activities

Activity	Application at Northparkes
5. Testing and evaluation of equipment/technology	*Partially undertaken.* Although the original design by Sandvik was conducted in Finland, the finished system was evaluated from a human centered perspective after it was deployed.
6. User error risk analysis in simulated mine emergency conditions	*Not undertaken.* The possible human error risks of the system in simulated mine conditions were not specifically examined before deployment. However, the automated system was deployed carefully in the change from manual control.
7. User manuals and training materials	*Partially undertaken.* Extensive training manuals and materials have not been formally developed, but an operator-centered training program does exist—consisting of paper-based training, shadowing an existing operator, and then operating the system whilst being supervised by an existing operator.
8. Integration of the equipment with other work systems	*Undertaken.* The deployment process integrated the automated system with other work systems at a mine site such as maintenance, communications, and vehicle data.
9. Identify unwanted events in actual operational conditions	*Undertaken.* Design-induced user error events in operational conditions are investigated—examples include using the laser scanners already in place for autonomous navigation to detect the proximity of the walls during manual operation and to convey this information to the operators through changes in color of the scanning information provided on the teleoperation assist window.
10. Feedback to designers for next generation of the equipment	*Undertaken.* The continuous presence of OEM expertise onsite allows a rapid feedback loop with designers.
11. HCD summary report	*Not undertaken.* Because the system was developed for just Northparkes, then this is not applicable. However, the case study presented here can be thought of a summary report of human-centered design issues for LHD automation design and deployment.

Overall, the analysis shows that most of these 11 human-centered design activities were successfully undertaken at Northparkes. Regularly involving both the operators and the developers of the automation was critical. Similarly, iterating the design, carefully evaluating changes, and submitting regular feedback to all stakeholders were important success factors.

5.2.3.1   Case Study Acknowledgements

We are grateful to Dr Joe Cronin, Geoff Butler, and the loader operators at Northparkes for their cooperation.

5.3   CASE STUDY 2: SANDVIK MINER BOLTER 650

5.3.1   Introduction

The Sandvik MB 650 is the current iteration of a continuous mining machine with integrated bolting rigs used for roadway development in underground coal mines. Roof and rib bolting can be undertaken simultaneously with cutting by miners standing on platforms on either side of the machine.

The first integrated miner bolter capable of simultaneous cutting and bolting (ABM 20) was deployed in Australia in 1991 (Price, 1991). The evolution of the safe design features of the Sandvik Miner Bolter since that time illustrates the potential for close relationships and extensive interactions between mining companies and equipment manufacturers to lead to innovations in equipment design that reduce injury risks and improve productivity.

Conventional "cut and flit" coal mining methods employ separate continuous mining machines and bolting machines. In this mining method, an entry is first cut using a continuous mining machine, then the mining machine is withdrawn and replaced by a bolting machine that installs the primary support. Integrating bolting and cutting functions in a single machine allows bolting to be undertaken immediately following cutting, very close to the face, which reduces the risk of roof fall. Reduced frequency of tramming of the mining machine also reduces exposure of pedestrians to collisions.

However, the need to frequently access the work platform of the integrated bolter miner leads to slip/trip risks during access/egress, and while working on the platform. Musculoskeletal injury risks are also associated with the manual handling required to place drill steels and bolts in the drill chuck during bolting, as well as handling mesh modules and vent tubes, and the mining machine power cable. Working in close proximity to the bolting rigs is associated with risks of interacting with the rotating drill steel, or the moving timber jack.

Other risks are exposure to dust and noise because miners stand on the platform while cutting is undertaken. There are also risks associated with struck by rock falling from roof or rib, and working in close proximity to numerous hydraulic hoses brings risks of fluid injection. Risk associated with performing maintenance tasks include being caught between moving parts, struck by hydraulic fluid, slip and fall, and overexertion caused by handling heavy loads in awkward postures.

The ABM 20 went through a range of design iterations including the addition of rib protection for the operator and enhanced temporary roof support, and on-board storage for bolting consumables. These changes were undertaken in close cooperation with users. For example, in a description of the development of a new longwall mine it was noted that ABM20 that

> Considerable redesign of the machine was undertaken with Voest Alpine to permit the machine to cut to 4.3 metres height... A work platform was designed for the operators

to stand 1.5 metres off the ground, be able to touch the roof and have all tools and supplies at their finger tips.... The platform was designed for operator convenience with a bolt box on either side of the machine located immediately behind the operator. Baskets were designed for chemicals, mono-rail fittings etc. Racks were fitted along the sides of the platform rails for spare drill steels and mono-rails. The ABM20 can carry sufficient supplies for 60 metres of roadway drivage. (Hayward, 1998: 231)

5.3.1.1 ABM 25 and ABM 25s

The ABM 25, introduced in 2003 at Newstan included a range of additional features. One innovative feature was the provision of a mesh handling system, which allowed multiple mesh sheets to be loaded by LHD over the conveyor tail, from where the sheets were hydraulically moved to sit in place on top of the mining machine. When required, each sheet was pulled forward and rotated into place on top of the temporary roof support when required in the bolting cycle. Although still requiring manual handling, this was a marked improvement over the previous practice of manually carrying each sheet and lifting it over the machine each time when required.

Other features of the initial version of the ABM 25 included height adjustable rib protection, high pressure water sprays, and face ventilation to reduce operator dust exposure.

Mining customers at this time were becoming acutely aware of the injury risks and the potential role of improved equipment design in reducing these risks. The design of the ABM 25 continued to evolve, driven by requirements of mining customers West Wallsend (Xstrata), Ulan (Xstrata), and Crinum (BMA). For example, platform handrails and ladders for access, and improved platform lighting and ventilation were introduced in 2005 at the instigation of West Wallsend (Figure 5.4).

BMA Crinum requested assistance from an ergonomist in determining appropriate dimensions for an ABM 25, taking into account user anthropometry and the variety of tasks to be performed (O'Sullivan, 2007). The outcome of the analysis

FIGURE 5.4 West Wallsend improvements included lighting on access, rubber floor mats, and handrails.

included the specification of a height adjustable platforms and stairway access for the ABM 25s introduced at Crinum and Ulan in 2006/2007. Push button control of bolting rigs was also first introduced at Crinum under the direction of Alan Bruce. Mr. Bruce subsequently became the product line manager for mechanical cutting at Sandvik where he influenced the subsequent development of the MB 650.

5.3.1.2 MB 650 Bolter Miner

The MB 650 bolter miner was introduced in 2009. The MB 650 featured a range of additional safe design features including: recessed platform lighting, semi-automated push button bolting controls, retractable platforms, hydraulically operated folding stairs, remote routine maintenance points, access for non-routine maintenance, canopy mesh locator, and onboard storage of mechanical isolation stops (Figure 5.5).

While the first push button bolting controls were introduced at Crinum in 2004, the design concept was been refined to current interface, which features two-handed fast speed control, grouping of related functions, and shielding to reduce the risk of inadvertent operation (Figure 5.6).

5.3.2 EDEEP PROCESS

As noted mentioned earlier, the Earth Moving Equipment Safety Round Table (EMESRT) engages with mining equipment manufacturers with the aim of accelerating improvements in the safe design of mining equipment. In 2011, EMESRT developed a process for the evaluation of equipment safe design intended particularly for use during the procurement process (EMESRT Design Evaluation for Equipment Procurement—EDEEP; Burgess-Limerick et al., 2012). The process involves identification of priority tasks and a task-based risk assessment process through which safety risks are evaluated and design solutions are identified (Figure 5.7). Of key importance, the process calls for the involvement of users throughout each step.

As noted previously in this book, Sandvik have adapted the process (the Sandvik EDEEP process) and have utilized this formal method of evaluating safe design features for drill rigs (Wester and Burgess-Limerick, 2015) and the MB 650.

In 2014, Sandvik assembled a team drawn from across the product safety and engineering functions in Zeltweg, Austria, where the equipment is designed and manufactured, along with others from Australia, and an experienced customer representative from BHP Billiton. The team systematically evaluated the consequences of potential unwanted events associated with 500 task steps undertaken by operators and maintainers of the MB 650 (Sandvik, n.d.).

Eleven task steps were selected for more detailed analysis:

1. Roof and rib bolting
2. Tramming to face
3. Cable bolting
4. Flitting/relocation
5. Loading bolting supplies
6. Loading mesh
7. Drill head torque test

FIGURE 5.5 MB 650 #181/182/204 (2010/2011).

8. Maintain spray nozzles at conveyor tunnel
9. Access platform
10. Inspect main frame and sump slide
11. Maintain picks

Detailed flow charts were developed and a task-based risk assessment was undertaken for each task step involving both experienced users and maintainers of the equipment. Current safe design features were evaluated and potential new design

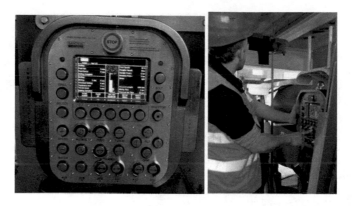

FIGURE 5.6   Semi-automated push-button roof bolting controls employed on MB 650.

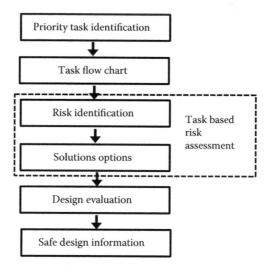

FIGURE 5.7   EDEEP process steps.

controls were identified in the areas of mesh-handling; sump slide inspection, spray nozzle maintenance, and boom stop handling.

The final product of the Sandvik EDEEP process is a "Sales Design Information" document that summarizes the outcomes of the task-based risk assessment process. The designers present during the process gained additional insight into realistic use-cases for the equipment, and the information obtained will also be utilized in operation and maintenance manuals; as well as customer training materials.

5.3.3   HCD ANALYSIS

Table 5.2 compares the Sandvik MB 650 redesign using the EDEEP process with the 11 human-centered design activities introduced previously in this book.

TABLE 5.2
Comparison of Sandvik MB 650 Redesign Case Study with the 11 Possible HCD Activities

Activity	Application to MB 650 Redesign
1. Understanding the need for the equipment, the likely users and the use environment.	*Undertaken.* The MB 650 was developed building on previous Sandvik integrated Miner Bolters in which end-user input was obtained to understand the need for the integrated equipment to replaced "cut and flit" mining methods, and to understand some of the safety risks such as the risk of a roof fall following cutting. Use scenarios for the integrated miner bolter equipment were identified- for example, linked to possible musculoskeletal injury, noise, and dust risks.
2. Task, incident and error analysis for the equipment	*Partially undertaken.* The likely tasks with the integrated Miner Bolter were explored during design—for example, the reduced task frequency of tramming of the mining machine reduced the exposure of pedestrians to collisions with the equipment
3. Participatory design with end users	*Undertaken.* The EDEEP process was used with the MB 650. It worked with end users to help design/evaluate new versions of equipment.
4. Application of relevant standards (e.g., ISO, ANSI) and HF data	*Unknown for the MB 650 development.*
5. Testing and evaluation of equipment/technology	*Undertaken.* The EDEEP process also evaluated the completed MB 650 in terms of operational and maintenance risks for 11 key tasks (roof and rib bolting). In addition to EDEEP with the MB 650, formative design was undertaken with earlier versions of Miner Bolters: for example, user anthropometry assessment with the ABM25
6. User error risk analysis in simulated mine emergency conditions	*Unknown for the MB 650 development.*
7. User manuals and training materials	*Undertaken.* The EDEEP process gave the designers additional insights for the development of operational and maintenance manuals and training materials.
8. Integration of the equipment with other work systems	*Unknown for the MB 650 process.* But older generations of Sandvik Miner Bolters undertook this activity. For example, the ABM25, platform handrails, and ladders for access were introduced in 2005 at West Wallsend.
9. Identify unwanted events in actual operational conditions	*As with activity 10, unknown for the MB 650 process.* But older generations of Sandvik Miner Bolters undertook this activity. The close relationships and extensive interactions between mining companies and the equipment manufacturer has previously resulted in innovations in equipment design for safety and productivity.

(Continued)

TABLE 5.2 (*Continued*)

Comparison of Sandvik MB 650 Redesign Case Study with the 11 Possible HCD Activities

Activity	Application to MB 650 Redesign
10. Feedback to designers for next generation of the equipment	*Undertaken.* The many iterations of integrated Miner Bolters by Sandvik is an example. Also, the MB 650 EDEEP process directly fed back issues to designers for the product's ongoing development.
11. HCD summary report	*Undertaken.* The EDEEP process with the MB 650 produced a Sandvik "Sales Design Information" that has similar information to a HCD summary report

Overall, the analysis shows that most of these 11 HCD activities were successfully undertaken. In particular, the evolution of the safe design features of the Sandvik MB 650 illustrates the value of close relationships and extensive interactions between mining companies and equipment manufacturers to lead to innovations in equipment design that can both reduce injury risks and improve productivity.

The use of EDEEP by Sandvik further expanded the use of HCD. This was primarily by means of end-user input into a task-based risk assessment for high priority operational and maintenance tasks for the MB 650. This shows that existing human-centered design tools can successfully be modified by OEMs to suit their precise purposes.

5.4 CASE STUDY 3: DUST SUPPRESSION HOPPER IMPLEMENTATION (CASE STUDY COWRITTEN WITH JESSICA MERRILL, NIOSH PMRD)

5.4.1 INTRODUCTION

5.4.1.1 Human-Centered Design of a Process

In addition to the mining equipment and technologies that have been the key focus of this book, human-centered design principles can be applied to process design. When designers think about a system—how a process is going to work—they need to think about what kinds of tasks and situations the process will expose workers to. The designer and mine site operational management needs to understand the relationship between the process and the worker, and how they together relate to:

- **The environment**: They need to consider the environmental conditions the process will create. Is this environment safe for workers? Does any step in the process change the environment?
- **Human error**: They need to think about human error potential. As noted in the previous chapter of this book, contemporary views in safety science regard human error as a consequence of sub-optimal systems design, so

designers must look for possible circumstances where human error may occur within their process. What errors can be made in this situation? How would these errors affect the user and the output of the process? If these errors may cause issues with the products, the environment, or for the users, what safeguards should be put in place to prevent these incidents from occurring, or make them less likely to occur?

- **Integration of technology into the process**: When determining what technology is going to be used in the process, they must consider what steps are necessary to use the technology (the process of using that technology). Are the tools used fitted to their users and uses within the process?

Therefore, a process needs to be designed with the user in mind. If the human components of a system are not fully considered, errors may be made that may have negative consequences for the worker or the equipment. Of course, problems that arise in the system should not be addressed as separate issues. Any solutions implemented should fix the root problems in the human-centered design of the process.

5.4.1.2 Background to the Case Study

An industrial sand processing and mining facility produces industrial sand for applications in the hydraulic fracturing, glass, and foundry industries. This facility had historically been using a telescoping spout system to load high purity silica sand into over-the-road, open top truck beds. The telescoping spout system could be lowered approximately 8–10 feet from the existing overhead-steel structures in order to reduce airborne dust exposure during loadout process, but operators may often not lower the spout because it was not always working properly (Figure 5.8).

FIGURE 5.8 Telescoping spout filling a truck bed (not lowered).

The open-top truck-beds loaded at this facility are typically separated into two compartments. In order to fill the truck beds, the front compartment is first filled; the driver stops the flow of material and moves the truck forward, and then starts the flow again until the rear compartment is filled. The telescoping spout would have to be raised and then lowered again in-between the two stages if the spout was used as designed.

Dust collection was engineered within the telescoping spout system, but unless the spout was descended properly the dust suppression mechanisms in the spout had minimal effectiveness. When the spout was not lowered, the material transfer distance was quite large, permitting the air to displace dust. This resulted in dust accumulation on surrounding surfaces and structures that required continuous housekeeping.

More importantly, the dust posed a significant health hazard for workers in the area, as it contained respirable silica. Exposure to respirable crystalline silica above the permissible exposure limit has been associated with silicosis, tuberculosis, chronic bronchitis, emphysema, and chronic renal disease, and is a known carcinogen (NIOSH, 2002). NIOSH has a recommended exposure limit of less than $50\,\mu g/m^3$ over an average day of up to 10 h (NIOSH, 1974). Exposure to dust, especially dust that contains silica, should be minimized.

The telescoping spout system was the only way to load over-the-road truck beds so, while maintenance activities were being performed on the spout customer, orders could not be fulfilled. Replacing the telescoping spout system with a dust suppression hopper was suggested to the plant through the company's internal engineering department. At the time, NIOSH PMRD was concurrently conducting studies on quantifying the effectiveness of dust suppression hoppers in similar applications (NIOSH, 2012). If proven effective, dust suppression hoppers would decrease dust concentrations that could affect worker health. Such dust suppression hoppers could also be useful for other industries, such as manufacturing, oil, and gas.

5.4.2 DUST SUPPRESSION HOPPER DESCRIPTION

Dust suppression hoppers are used to reduce the number of particles released into the environment when loading loose material into a storage container or pile. Material is loaded through the top of the cone-shaped hopper, and is released at the bottom. A small amount of "natural agitation" inside the cone removes much of the air within the load, allowing for the discharge from the cone to be more concentrated and flow more evenly. Since less air is permitted inside the cone, the amount of material being displaced from the direction of flow is reduced, therefore reducing the amount of dust. This dust suppression hopper uses no electrical power and has no moving parts on the inside, so it is advertised to require little to no maintenance.

Prior to the installation of the dust suppression hopper, the plant conducted a basic experiment to verify and quantify the hopper's effectiveness. If the hopper proved to be effective, the engineering department would support a wider adoption of the hoppers in other applications within the company. A quantitative evaluation was conducted to compare dust levels from before and after replacing the telescoping spout with the dust suppression hopper. A personal DataRAM (pDR-1000AN) was used to measure the concentration of airborne particulate matter (dust) approximately 8 in. below the top of the truck bed during filling for both pieces of equipment. The pDR was suspended by a rope from the loading surface into the truck bed (Figure 5.9). Data were collected for the entire filling process for two trucks for the telescoping spout and the dust suppression hopper conditions.

The study found that the average dust concentration level was greatly reduced when using the dust suppression hopper instead of the telescoping spout. The reduction of airborne dust also decreased the amount of dust that settled onto the surrounding surfaces. This settled dust could be re-entrained any time, causing additional exposures to respirable silica. For example, settled dust near the telescoping spout could be agitated and reintroduced during maintenance activities, exposing the maintenance workers to harmful dust. The new equipment significantly limited worker exposure to respirable crystalline silica, thus creating a safer environment.

An additional benefit was that after the installation of the dust suppression hopper, less maintenance was required. For at several months following the installation of the new system, no maintenance was needed on the hopper. Significant additional cost savings were observed after considering the amount of man-hours, equipment

FIGURE 5.9 pDR collecting data for telescoping spout truck loading.

needed, and downtime associated with repairs, service, and routine physical maintenance consumed with the previous telescoping spout loading system.

Overall, the process for loading trucks with the dust suppression hopper suits the
user better than the previous telescoping spout technology. The dust suppression
hopper was easier to use, required less maintenance, and created a cleaner, safer
environment for the employees.

5.4.3   HCD Analysis

Table 5.3 compares the dust suppression hopper implementation with the 11 human-
centered design activities introduced previously in the book.

Identifying the parts of the telescoping spout that were difficult or dangerous for
the users, the plant was able to identify a new technology that did consider the user
in its design, and created a process that was more human centered. A process must
maintain a safe environment, minimize error potential, and integrate tools that are
easy to use correctly in order to be user-friendly. These aspects should be at the fore-
front of a design or redesign of a process.

5.4.4   Case Study Conclusion

Human centered issues may or may not be the reason for redesign of a product or
process, but HCD principles should be applied in redesign for the same reasons they
should ideally be considered in initial design. As seen in this dust hopper example, applying HCD principles in redesign can help produce a system that improves
to worker health and safety issues, and reduces the occurrence of human error.
Redesigns that explicitly consider the user can create a more usable process.

5.5   CASE STUDY 4: INTERFACE DESIGN FOR HAUL
TRUCK PROXIMITY ADVISORY SYSTEMS

5.5.1   Introduction

At least 24 fatalities have occurred at surface mines as a consequence of a haul
truck colliding with another vehicle (Burgess-Limerick, 2017). Eight of the fatalities
occurred when a haul truck was driven over a stationary light vehicle. Another eight
occurred at intersections when a haul truck collided with a moving light vehicle. For
the majority of fatalities (79%), a causal factor was the haul truck driver having an
inaccurate understanding of the location and movements of other vehicles in the proximity of the driver's truck, commonly referred to as a "loss of situation awareness."

Original equipment manufacturers and after-market providers have offered a
range of proximity advisory systems to assist truck drivers maintain accurate situation awareness by providing information regarding the proximity and movement
of other vehicles. Regardless of the means by which such information is obtained,
the information is communicated to the truck driver through auditory and visual
interfaces.

TABLE 5.3

Comparison of the Dust Suppression Hopper Case Study with the 11 HCD Activities

Activity	Dust Suppression Hopper Application
1. Understanding the need for the equipment, the likely users and the use environment.	*Undertaken.* Identified user issues including: difficulty of use, worker health hazards, frequent breakdowns that required maintenance, and housekeeping needs. Evaluated use of technology to determine that workers were not using the telescoping spout as intended. In addition, Sioux Steel representatives evaluated intended use before implementation.
2. Task, incident and error analysis for the equipment	*Undertaken.* New technology was determined not to have the issues that the telescoping spout posed. Compatibility with the existing loadout structure and product flow were also considered.
3. Participatory design with end users	*Undertaken.* Operators helped identify user issues by reporting problems as they occurred.
4. Application of relevant standards (e.g., ISO, ANSI) and HF data	*Not undertaken.*
5. Testing and evaluation of equipment/ technology	*Undertaken.* Measured dust levels for telescoping spout and dust suppression hopper with pDR.
6. User error risk analysis in simulated mine emergency conditions	*Undertaken.* Dust suppression hopper at this facility served as a pilot study for possible application at other company sites. Data collected and problems identified would have improved implementation at other facilities.
7. User manuals and training materials	*Undertaken.* Operators and maintenance workers were trained on using the dust suppression hopper after its installation.
8. Integration of the equipment with other work systems	*Partially undertaken.* No formal evaluation of user acceptance was completed, but management received positive feedback from end users.
9. Identify unwanted events in actual operational conditions	*Partially undertaken.* No formal evaluation of failure modes completed, but no issues were reported by the operators.
10. Feedback to designers for next generation of the equipment	*Undertaken.* Dust reduction evaluation study was documented and presented.
11. HCD summary report	*Not undertaken.* The company did not complete an HCD summary report, but this case study can serve as the report for this application.

For any such proximity awareness technology to be effective in preventing collisions, the following steps must be undertaken accurately and quickly by the truck driver:

- Detection—the driver attends to the proximity awareness system interface.
- Perception—the information provided is interpreted to provide an accurate understanding of the current situation.
- Prediction—the probability of a future collision is predicted.
- Decision—a decision is made regarding what action should be taken.

Errors, or delays, at any of these stages may lead to failure of the proximity awareness system to prevent a collision. The design of the interface by which information about the proximity of other vehicles is provided is consequently likely to play a critical role in determining the effectiveness of any proximity advisory system as a control measure to prevent collisions.

Warning tones are typically provided to attract attention to the proximity awareness interface. Variation in timing, pitch or volume of sounds may also be used to convey information about the situation. Auditory information is typically accompanied by a visual display to enhance a driver's ability to maintain situation awareness and make accurate and timely decisions to maintain vehicle separation (Horberry et al., 2016).

The visual interfaces that have been designed for use as part of proximity advisory systems intended for use on mining haul-trucks may be broadly divided into those that provide an alarm accompanied by an indication of the direction of vehicle giving rise to the alarm, and those that provide additional information about the identity, location, and state of other vehicles with respect the operator's truck. An optimal proximity awareness interface is one that alerts the operator when, and only when, attention to the interface is required; allows the operator to quickly and accurately understand the current and likely future locations of other vehicles with respect to the driver's own vehicle; and consequently, allows the driver to make appropriate and timely adjustments to throttle and/or brake and steering to ensure adequate separation between vehicles is maintained.

The aim of the project described in this case study was to examine the consequences of different visual interfaces when drivers are presented with a range of potential collision scenarios within a mining haul truck simulator. These data, in combination with best practice information obtained from other industries, were used to derive guidelines for the design of interfaces designed to convey proximity advisory information to haul truck drivers.

5.5.2 HAUL TRUCK SIMULATOR

A haul truck simulator designed for use in the resources sector by the company 5DT for operator training was adapted as a unique research tool. The simulator features a six-degree-of-freedom motion platform, realistic haul truck control layout, and three projector screens (Figure 5.10). Collaboration with 5DT has allowed access to the underlying simulation software and the possibility to create standardized collision

FIGURE 5.10　5DT haul truck simulator located in the Centre for Sensorimotor Performance, School of Human Movement and Nutrition Sciences, The University of Queensland, which was employed in the experiments (see www.5dt.com/).

provocative scenarios, and the recording of data describing simulator operator behavior for subsequent analysis.

Participants drove the simulated haul truck around a circuit illustrated in Figure 5.11. Each trial consisted of driving from a park-up area around three corners (1–3) to the top of a ramp (corner 4) then driving down the ramp to an excavator. Following loading, the participants navigated the simulated haul truck to corner 5, back up the ramp, and then around three corners (7–9) before reaching a dump point where the trial was ended. Each circuit took about 20 min. The participants' instructions were to navigate the simulated haul truck around the circuit as quickly as possible while not exceeding the speed limit of 40 km/h, and to avoid colliding with any other vehicles.

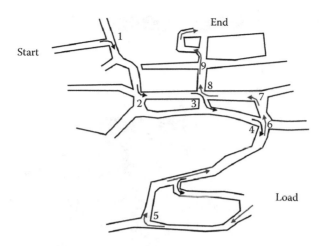

FIGURE 5.11   Circuit driven during each experimental trial.

The simulation included four other haul-trucks autonomously traveling a similar circuit. Six light vehicles were also parked, or traveling around the circuit during each trial. The position and travel paths of these vehicles varied between trials. The locations and travel paths of the light vehicles were controlled via a scripting language (Lua).

Nine different scenarios were scripted. In each scenario, light vehicles were programmed to create a situation with potential for a collision to occur at two of the corners. For example, in two of the nine scenarios, as the simulated haul-truck approached corner 1, preparing to turn right, a light vehicle travelled from right to left passing in front the simulated truck and obliging the participant controlling the simulated haul truck to take action to avoid a collision. Similar collision potential events were coded to occur in two scenarios at corners 2–7 and 9.

In two scenarios, a stationary light vehicle was located at corner 5 requiring the truck driver to ensure the truck did not cut the corner. In two of the nine scenarios a stationary light vehicle was also located immediately in front of the simulated haul truck at the commencement of the trial. The collision event coded for corner 6 was subsequently found to be ineffective because of the low speed of the fully loaded haul-truck traveling up the ramp obviated the need for any action to be taken by the truck driver to avoid the collision. This potential event was omitted from subsequent analysis. No collision potential events were coded for corner 8.

The experimental tasks were undertaken in three different proximity advisory interface conditions. In a "control" condition, no proximity advisory information was provided and drivers relied on line-of-sight information about the location of other vehicles in the simulation to avoid collisions. In the remaining two proximity advisory interface conditions, additional information about the location of other vehicles in the simulation was provided via warning tones and a visual interface. Information about the identity, location, and proximity of other vehicles in the simulation was continuously transmitted wirelessly from the haul truck simulator to a Samsung tablet located on the truck dash-board via a dedicated wireless network connection.

Identical auditory information was included in both proximity advisory interface conditions: a low pitch initial alert tone occurred when another truck or light vehicle first approached within 150 m, following by two further warning tones of successively higher pitch as the proximity to another vehicle became less than 100 and 80 m, respectively. Visual information was provided in one of two ways, either a "ring" interface, or a "schematic" interface (Figure 5.12). The "ring" interface condition consisted of a ring of simulated LEDs on the tablet that became illuminated indicating the direction of approach and distance range of other vehicles in proximity to the participants' truck. The simulated LED were illuminated yellow when another vehicle approached with 150 m, red if the vehicle approached within 100 m, and flashing red if within 80 m. The "schematic" interface provided a continuously updated display of the location of other vehicles. The schematic interface provided additional information over the ring interface in that the relative velocity of approach of surrounding vehicles was available to the operator as the display was continuously updated. The type of approaching vehicle (truck or light vehicle) was also indicated by the schematic interface.

The instantaneous location of all vehicles in the simulation, and the state of vehicle controls were transmitted from the 5DT simulator and captured at 10 Hz on a separate laptop. Simultaneous video recordings were made of both the driver's view of the simulation and an overview of the simulation, including representation of the driver's controls. Combining these data sources allowed both qualitative and quantitative aspects of the simulation to be subsequently assessed and analyzed. Analysis of the vehicle location data was undertaken via custom MATLAB® scripts. A range of measures were calculated for each potential collision situation including the local minimum of the Time-To-Collision between the participant's truck and the light vehicle at corners where collision vehicles were coded, the maximum braking force applied during the corner approach, and the minimum speed of approach to the each corner.

(a)   (b)

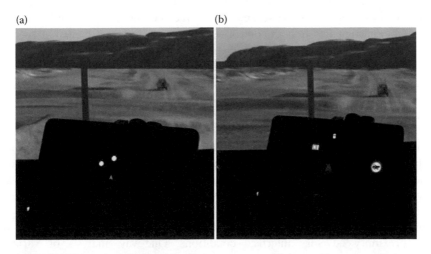

FIGURE 5.12   Ring (a) and Schematic (b) interfaces indicating the presence of another truck (unseen) and a stationary light vehicle.

5.5.3 Experiment 1

5.5.3.1 Participants and Experimental Design

Thirty-six participants (17 female, 19 male; aged 18–56 years, mean 27 years) were recruited. All were currently licensed light vehicle drivers without truck driving experience. Each participant performed two training circuits during an initial session, followed by 18 experimental trials undertaken over three separate days. A fully-between experimental design was employed—an approximately equal number of female and male participants were randomly assigned to perform the experimental trials in one of the three proximity advisory interface conditions only. Each participant performed each of the nine scenarios twice during the 18 trials in pseudo-random order. The order of scenarios was balanced across the three proximity advisory conditions. Each participant was consequently presented with each collision potential event four times across the three days of testing. Median values for each participant were calculated for the measures describing cornering behavior at corners 1–5, 7, and 9 for those scenarios during which collision potential events occurred, and for corner 8 for all scenarios as an indication of cornering behavior in the absence of other vehicles creating a potential collision hazard.

5.5.3.2 Results

The data obtained for each participant from seven corners during scenarios at which potential collision events occurred were combined for statistical analysis. Figure 5.13 illustrates the mean and 95% confidence intervals of these combined data, as well as the 95% confidence intervals of the pair-wise differences between conditions for each measure. One-way analysis of variance indicates that the effect of interface condition on the Time-To-Collision local minimum and the minimum cornering speed was statistically significant ($F[2,33]=4.522$, $p=0.0184$ and $F[2,33]=14.43$, $p < 0.0001$, respectively) but that the effect on maximum braking force was not ($F[2,33]=2.34$, $p=0.1121$). Pairwise comparisons indicated that the cornering behavior of participants in the schematic interface condition was significantly different from participants in both control and ring interface conditions. On average, across these seven corners, these participants exhibited higher minimum Time-To-Collision and higher minimum cornering speed. No significant differences were observed between the ring and control conditions when the data were combined across the seven corners.

A stationary light vehicle was located in front of the participants' simulated haul truck on 4 of the 18 trials each participant performed. Figure 5.14 illustrates the situation, and the participants' view from the cab in this situation, which includes the light vehicle flag in the lower right (circled). The video records of the trials were inspected to determine whether the participants' first movement of the truck was forward (a collision) or reverse, and expressed as a probability of collision for each participant.

Figure 5.15 illustrates the mean probability of collision for participants in the three proximity advisory interface conditions. One-way analysis of variance indicated that the effect of interface condition was significant ($F[2,33]=12.36$, $p < 0.0001$). Pair-wise comparison indicated that both ring and schematic conditions

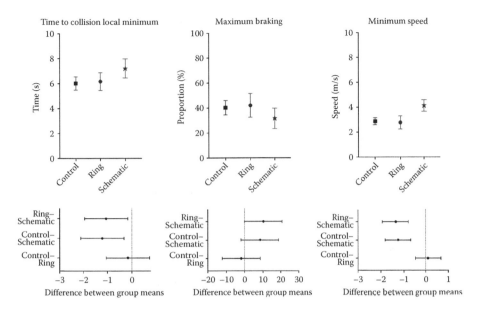

FIGURE 5.13 Measures describing cornering behavior when participants were presented with potential collision collated across seven corners (1–5, 7 and 9). Error bars indicate 95% confidence limits. The lower panel indicates the magnitude of pair-wise differences between conditions and the 95% confidence intervals of these differences. Confidence intervals that do not include 0 may be considered to be statistically significant.

FIGURE 5.14 The situation encountered by participants on 4 of the 18 trials undertaken.

were associated with a lower probability of collision than the control condition, but that there was no significant difference between ring and schematic conditions.

5.5.4 EXPERIMENT 2

5.5.4.1 Participants and Experimental Design

Similar data were obtained from 11 experienced haul-truck drivers. The drivers (10 male, 1 female, aged 30–42 years, mean 37 years) had between 3 and 12 years' experience operating surface mine haul trucks (mean = 7 years). Each driver attended

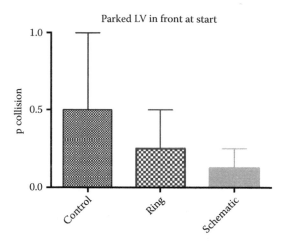

FIGURE 5.15 Probability of collision when a stationary light vehicle was located in front of their truck at the start of the trial. Error bars indicate 95% confidence limits.

the university campus for 1 day. Two training circuits were completed, followed by 18 experimental trials. In this case, a fully-within subjects design was employed and each driver performed six of the scenarios from experiment 1 in each of the three proximity advisory control interface condition. The scenarios were presented in randomized and balanced order. The scenarios included collision potential events at corners 1–5, 7, and 9. Conversational interviews were undertaken with each truck driver to obtain their views on the relative merits of the different interface conditions.

5.5.4.2 Results

Figure 5.16 illustrates the average minimum Time-To-Collision and the minimum cornering speed calculated across truck drivers for the corners at which collision potential events occurred. Although the results were qualitatively similar to those obtained from novice drivers, the differences between conditions were much smaller and were not statistically significant.

The experienced drivers were observed to drive the simulated haul truck far more conservatively than the participants in Experiment 1 and this was evident in much reduced minimum cornering speed at a corner where light vehicles were never encountered. When interviewed, the experienced drivers reported a clear preference for the schematic display and commented favorably on the provision of additional information through this means. The drivers were also very consistent in indicating that being alerted to the presence of other vehicles that were not a collision risk was counterproductive and highly undesirable.

5.5.5 Case Study Discussion

The data collected from novice truck drivers provide clear evidence that visual interface design is likely to influence the effectiveness of proximity advisory control systems, and that the additional information available in the Schematic interface was

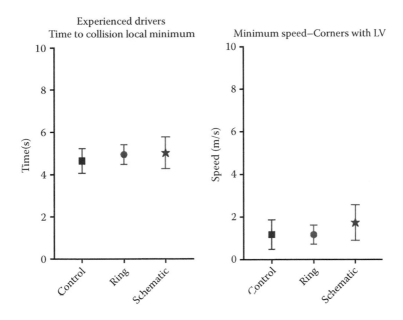

FIGURE 5.16 Experienced drivers' cornering behavior at corners where collision potential events occurred. Error bars indicate 95% confidence limits.

utilized to reduce collision risk and reduce travel time. When faced with situations with collision potential, novice driver participants assigned to the Schematic interface condition used the additional information available to adjust their speed earlier than participants assigned to either the Ring interface, or Control conditions, resulting in a higher minimum Time-To-Collision, less extreme braking, and a higher minimum cornering speed. Although some differences were observed between participants assigned to the Ring and Control conditions at individual corners, these differences were not statistically significant.

The differences in cornering behavior exhibited by novice driver participants were even more marked when no other vehicles were located in proximity to the simulated haul truck. In this situation it appears that the novice driver participants assigned to both proximity advisory interface conditions utilized the information indicating the absence of any other vehicles in proximity of the corner to minimize braking and to corner at a higher speed. This was particularly so for participants assigned to the Schematic interface condition. These effects led to reduced total travel time to complete the circuit for participants assigned to the Schematic interface condition.

Clear differences were also evident when novice participants were placed in situations in which a stationary light vehicle was located in front of the simulated truck at the beginning of a trial. Novice truck driver participants assigned either proximity interface condition were significantly less likely to collide with the light vehicle. Average collision probability in this situation was lower in the Schematic condition than the Ring interface condition, although this difference was not statistically significant.

In contrast, although small differences consistent with those already noted were observed in the data obtained from experienced haul truck drivers performing a related experiment, the effects were not statistically significant. There are several reasons that may explain why smaller differences were found. The experienced drivers were observed to generally drive more conservatively. At those corners where light vehicles created collision potential, the experienced drivers were more likely to come to a complete stop. Consequently, although the minimum Time-To-Collision was slightly higher in conditions where proximity advisory information was provided, and the minimum cornering speed was slightly higher when this information was provided via the Schematic interface, the experienced drivers' average minimum cornering speeds at corners where light vehicles were present was lower that the corresponding speeds calculated for novice participants when the proximity advisory information was provided. Despite the absence of any signage from the simulation, the experienced drivers appeared to treat many of the corners as if stop signs were present. The experienced drivers were seemingly disinclined to drive through intersections without stopping, even in situations where the proximity advisory system present indicated the absence of other vehicles in proximity to the simulated haul truck.

It may also be that within-subjects experimental design employed in the second experiment reduced the potential potency of the interface condition manipulation. Rather than being assigned to one condition and performing 18 trials (approximately 6 h of driving) in that condition as in the first experiment, the experienced participants each performed six trials in each of three interface conditions, with the order of presentation of trials assigned such that each group of three trials contained one trial in each interface condition. It may be that this reduced the likelihood of the experienced drivers becoming accustomed to, and trusting, the proximity advisory system in the conditions where the information was provided.

When the drivers were questioned about their views on the provision of the proximity advisory information, and the different visual interfaces, there was near-unanimous agreement that the information was potentially useful, and that the Schematic interface was preferred. The drivers' primary concerns about the proximity advisory system was that the alarm tones intended to draw the drivers' attention to the interface should only occur when a genuine risk of collision was present. For example, the system should not alarm for a light vehicle traveling in the opposite direction to the haul truck on the other side of the haul road.

5.5.6 DESIGN GUIDELINES

The results of these experiments, in conjunction with the review of best practice in other domains and related standards (Horberry et al., 2016), provide a basis for deriving the following design guidelines for proximity advisory systems for use in haul trucks.

- Proximity advisory systems should provide information to haul truck drivers via both auditory and visual displays.
- Visual displays should be located within the drivers' primary display location as defined by EN894-4 (that is, between 5° and 40° below horizontal

eye height, and less than 35° laterally) and positioned to avoid reducing the driver's external visibility.

- Proximity advisory system visual displays should provide information regarding the identity, position, and relative velocity of other vehicles in proximity to the driver's truck.
- Auditory collision warnings should sound when, and only when, the relative velocity of vehicles in proximity to the driver's haul truck indicates an imminent potential collision.
- The Time-To-Collision threshold chosen for warnings should vary as a function of truck speed, and allow for driver reaction and movement times (see additional design issues later in this chapter).
- To avoid false positive alarms, the accuracy of predicted collisions should be enhanced by roadway and elevation information.
- An Auditory Icon, such as a car horn, should be provided for imminent collision information. Earcons, such as a musical tone, may be used to convey cautionary information if multiple warning stages are utilized (see additional design issues later in this chapter).
- An auditory warning should not be presented more than three times per collision potential incident. These repetitions should occur in immediate succession.
- If a speech-based warning is provided to convey an imminent collision threat, the message should be kept to a single word e.g., "Brake." A speech warning should not be preceded by an alerting tone. False alarms have high annoyance potential.
- An auditory collision alarm should provide directional information, orienting the driver to look in the direction of the hazard.
- The amplitude of auditory signals should be 20–30 dB above ambient noise levels. In-vehicle systems that provide concurrent audio should be muted during the presentation of an auditory warning.
- If multiple hazards occur simultaneously, the driver should be provided with auditory (and visual) warning of the highest priority hazard, while information regarding lower priority hazards is only provided visually.

Some details of the optimal design of such interfaces require further research. These include:

- The number of warning stages—guidelines from other domains suggest that a single stage alarm may appropriate if the primary aim of the system is warn a distracted or fatigued driver of an impending collision; a multiple stage alarm may be preferred if the aim is to promote safer headway distances, or overcome restricted visibility;
- False-alarm rate—while the false-alarm rate should be minimized, what constitutes an acceptable false-alarm rate has not been determined for the haul truck context (less than once per week has been suggested as appropriate in other contexts);

- The Time-To-Collision threshold chosen for auditory warnings should allow for reaction and movement times, however these data are not currently available for the surface mine haul truck situation, taking into account potential fatigue and time-of-day effects;
- Training—the extent and nature of training required for haul truck drivers to become competent in utilizing the information provided by proximity advisory systems has not been determined.

5.5.7   HCD Analysis

Table 5.4 compares the interface design for Haul truck Proximity Advisory Systems with the 11 human-centered design activities introduced previously in the book.

This shows that the majority of the 11 activities were undertaken. This includes careful testing, haul truck collision analysis, experienced operator input, the use of standards and best HF practice, and the development of design guidelines.

5.5.8   Case Study Conclusion

Visual interface design influences the effectiveness of proximity advisory systems in both safety and performance terms. When additional information was available in the visual interface, it was utilized by novice participants to reduce collision risk and braking force, and decrease travel time. Design guidelines for proximity advisory systems for use in surface mining haul trucks are provided, based on the results of these experiments in conjunction with the review of best practice in other domains and related standards. A series of issues requiring further research were also identified, including the optimal number of warning stages, acceptable false-alarm rate, optimal Time-To-Collision threshold, and the extent and nature of training required to achieve competence in the use of proximity advisory systems.

5.5.8.1   Case Study Acknowledgements

The project was motivated by discussions with the Earth Moving Equipment Safety Round Table advisory committee and undertaken with funding from the Australian Coal Association Research Program (project C24028). The research leader was Professor Robin Burgess-Limerick. The experiments were conducted in the Centre for Sensorimotor Performance, School of Human Movement and Nutrition Sciences, The University of Queensland with the assistance of A/Prof Guy Wallis. The willingness of the participants to undertake the lengthy experiment is appreciated, as is the contribution made by Peabody Energy Australia to recompense the experienced drivers. Programming for the interfaces was undertaken by David Orchansky. Ben Burgess-Limerick created the scenarios in Lua and wrote MATLAB code to analyze truck driver cornering behavior. Danellie Lynas spent many, many, hours in the dark running the experiments. Nicholas Schepers (5DT) provided timely assistance to keep the simulator running. We are grateful to John Brett and Campbell Davidson (ACARP) and the industry monitors (Tony Egan, Allan Gordon, Ivan Heron, Allan Miskin, Gavin White, and Kane Usher) for their assistance.

TABLE 5.4
Comparison of the Interface Design Work for Haul truck Proximity Advisory Systems with the 11 HCD Activities

Activity	Interface Design for Haul truck Proximity Advisory Systems Application
1. Understanding the need for the equipment, the likely users and the use environment.	*Undertaken.* Identified user issues including reviewing previous designs of proximity advisory systems, undertaking site visits and operator interviews and receiving substantial guidance from experienced members of the Earth Moving Equipment Safety Round Table.
2. Task, incident and error analysis for the equipment	*Undertaken.* The research team undertook incident analysis to better understand causal factors in haul truck collisions.
3. Participatory design with end users	*Partially Undertaken.* The experienced haul-truck drivers used in the second study gave their input into which design they considered to be most effective, an acceptable false alarm rate and situations where the alert should not be activated.
4. Application of relevant standards (e.g., ISO, ANSI) and HF data	*Undertaken.* A review of best practice in other domains (e.g., automotive) as well as reviewing guidelines and standards was completed. For example, EN894-4 for visual display location.
5. Testing and evaluation of equipment/technology	*Undertaken.* The two experiments reported in the case study involved extensive user testing.
6. User error risk analysis in simulated mine emergency conditions	*Undertaken.* The two experiments reported in the case study involved simulated mine collisions involving the haul-truck and other vehicles likely to be present at the mine site.
7. User manuals and training materials	*Partially undertaken.* Operator training and competency with the proximity systems was explicitly recommended as an area for further investigation.
8. Integration of the equipment with other work systems	*Not undertaken.* Not within the current scope of this research, development, and testing project.
9. Identify unwanted events in actual operational conditions	*Not undertaken.* Not within the current scope of this research, development, and testing project.
10. Feedback to designers for next generation of the equipment	*Undertaken.* The outcomes of the research are being disseminated to the mining industry and technology manufacturers and designers.
11. HCD summary report	*Partially undertaken.* It did not explicitly complete an HCD summary report, as this was not within the scope of this research, development and testing project. However, the test results and design guidelines contain key information for a HCD summary report.

6 What Now for Mining HCD?

6.1 HCD ANALYSIS: WHAT WORKS?

The case studies in the previous chapter, as well as the shorter examples throughout the book, demonstrate that HCD can be an effective process in the development of fit-for-purpose mining equipment, systems, processes, and new technologies. It is applicable both to traditional mining equipment and to new technologies/systems.

HCD can also be a very flexible process. As the case studies and other examples show, there is no single way to conduct human-centered design. Instead, a continual focus on users and their tasks throughout the iterative design process is the essence of HCD.

Twelve specific success factors and teachings from the case studies and examples presented in the book are:

1. Start early: get HCD involved from the outset of a project. Although human factors and/or usability engineering are often introduced only after the design is largely fixed, the evidence clearly shows that it is generally cost effective to consider human element issues as early as possible.
2. Actively involve the people (usually this means operators and maintenance personnel) who will be the ultimate end users or who are otherwise somehow impacted by equipment/automation changes.
3. Maintain constant two-way communication between end users and designers. Ideally, designers should regularly visit mine sites to better understand the context of use, the tasks, and the end users of the equipment. Similarly, it is always very beneficial to have end users involved in the iterative design process.
4. By better understanding user requirements, provide operators with the information they need. Conversely, avoid providing information the operators do not need (e.g., irrelevant alarms).
5. Where possible, provide the operators with flexibility in ways of working. This is particularly important with complex, automated mining systems where different work strategies could be employed.
6. Consider operator training and competency needs early in the design and development process, as well as how competency will be assessed.
7. Empower operators to take action. This may include reporting design deficiencies or redesign suggestions to designers/manufacturers, or in some instances even making adjustments to the system themselves (as seen in the Northparkes case study, in which operators developed a camera cleaning system).

8. Take advantage of new possibilities with automated systems (e.g., optimal extraction patterns). For this, end user feedback, participatory-design approaches, and a flexible way of working will all be of great assistance.

9. Use human-centered design summary reports as a means to demonstrate how human element issues have been addressed in design. Mine sites, mining corporate procurement, and regulators can assist here by requesting such reports.

10. Build on past success: successful equipment design may develop iterative over several versions. This applies as much to traditional mining equipment such as haul trucks and bulldozers as it does to automated systems.

11. Use a task-based risk assessment process like EDEEP to evaluate equipment safe design, particularly for procurement. Similarly, use the other HCD tools presented throughout this book, such as OMAT, task analysis, usability evaluations, and in-depth interview techniques such as the critical-decision method.

12. Disseminate learnings, approaches, and successes to the wider minerals industry for others to use. In some cases, this might also include disseminating what did not work and explaining the reasons for the failures.

6.2 AUTOMATED MINING AND HCD

In additional to traditional mining equipment, this book has also focused on new technology and automation. Human-centered design can be applied at many different "levels" of automation in mining. There are several automation taxonomies for vehicle automation (such as NHTSA, 2013), but perhaps the most commonly-used is the J3016 standard developed by the Society of Automotive Engineers (SAE) (2014). The SAE classification system for automated road vehicles is also largely applicable to mining. This classification taxonomy identifies six different levels of vehicle automation.

- **Level 0 (no automation)**: The driver/operator is completely in sole control of the primary vehicle controls—brake, steering, throttle, and power—at all times and in all situations.
- **Level 1 (driver assistance)**: Individual vehicle controls are automated, such as electronic stability control; however, the driver is still in control of the vehicle.
- **Level 2 (partial automation)**: Some automation of braking, steering, and acceleration, such as adaptive cruise control, but the driver must have hands on the wheel at all times as well as actively monitoring the driving environment.
- **Level 3 (conditional automation)**: All aspects of driving are automated. The driver can give control of all safety-critical functions in most conditions, but must be ready to retake control.
- **Level 4 (high automation)**: The vehicle performs all safety-critical functions. The driver is not expected to control the vehicle at any time. However,

the vehicle is still equipped with manual controls, such as a steering wheel for instances of manual driving.

*   **Level 5 (full automation)**: The automated vehicle performs all functions and is not even equipped with a steering wheel or pedals. Manual driving is therefore never required, and in fact is not possible.

To give further examples of functions from the automotive domain, Level 1 automation includes conventional cruise control and ABS. Level 2 automation can control vehicle speed, lateral position in the driving lane, and distance from objects in front of the vehicle. Level 3 includes the automated vehicle changing lanes and making turns. In addition, automated monitoring of environment starts at Level 3 but the driver should stay in the control loop due to responsibility in the case of emergency. In Level 4, it is not necessary for the driver to stay in the vehicle control loop. Finally, in Level 5 it is not even possible for the driver to stay in the control loop due to a vehicle having no manual controls.

The HCD examples described in this book include no automation, assistance, and/or partial automation (as seen in the work of Cloete and Horberry (2014) for shovel assist technology) through to conditional and/or high automation for mining operations (as seen in the Northparkes underground automation case study where operators located in a surface control room load ore at draw point using manual tele-operated control; the loader is then switched to automated mode to travel to the bin where the ore is dumped autonomously; the loader then returns to the next draw point selected autonomously).

Given the increasing use of automation and new technology in the minerals industry, it is strongly recommended that further HCD work is undertaken with different levels of mining automation to effectively integrate humans and technical systems. The approach outlined in this document of human-centered design activities at different stages of the life-cycle could be an effective template for this work with new mining technologies and automation.

6.3   HOW DOES HUMAN-CENTERED DESIGN RELATE TO HUMAN SYSTEMS INTEGRATION?

Originally conceived in the context of defense procurement, human systems integration (HSI) is the broad overarching framework applied within defense and increasingly within other industries to ensure that total system performance is optimized and total operating minimizing costs are minimized. For example, US DoD (2008) instruction 5000.02 requires a procurement program manager:

> ... to have a plan for HSI in place early in the acquisition process to optimize total system performance, minimize total ownership costs, and ensure that the system is built to accommodate the characteristics of the user population that will operate, maintain, and support the system.

Booher (2003) defined HSI as the process of integrating the domains of human factors engineering, system safety, training, personnel, manpower (crewing), health hazards, and survivability into each stage of the defense-systems-capability life cycle (needs, requirements, acquisition, service, and disposal) where:

- *Human factors engineering* is defined as the systematic application of information about human capabilities, limitations, characteristics, behavior, and motivation to the design of equipment, facilities, systems, and environments
- *Systems safety* is the process of minimizing safety and health risks through identifying, assessing, and controlling hazards associated with the system
- *Manpower* (crewing) refers to the number of persons required to operate, maintain, sustain, and provide training for systems
- *Personnel* refers to the aptitudes, experience, and other personal characteristics required
- *Training* refers to the instruction and training required to fulfill the person's role in the system,
- *Health hazards* refer to conditions inherent in operation and use of a system that may cause death, injury, illness, disability or reduce the performance of personnel, and
- *Survivability* refers to the characteristics of a system in order to reduce fratricide, the probability of being attacked and war fighter injury.

Human-centered design, as described in this book in the context of the minerals industry, refers to the processes by which the goals of human systems integration are achieved; with a particular focus on the domains of human factors engineering, systems safety, health hazards, and training. Considering the broader context of HSI directs a designer's attention to consider that design decisions in one domain or dimension will very likely have consequences for other domains, introducing a requirement for designers to consider trade-offs between different domains (Durso et al., 2015). Equally, understanding this broader context is important as it shows that safety and performance in complex mining systems can impacted by decisions and actions made at all levels of the system, not just by operators and maintainers working with the equipment at the sharp end.

6.4   OVERALL CONCLUSIONS

This book has shown that the application of HCD can be highly beneficial as part of the design of safe and effective mining equipment and new technology. It is anticipated that the educational guide for mining human-centered design presented in Chapter 4 will better explain how different HCD activities can be undertaken, and when they can be used in the equipment lifecycle. Equally, the detailed case studies presented in Chapter 5 and the examples provided through the whole book show how mining human-centered design has been successfully undertaken in the past.

However, as has been discussed earlier, mining HCD is still a very small field, and putting the human at the center of the design process is not common for mining equipment and new technology. To further expand the use of human-centered design in mining so that it becomes routine, it is perhaps not enough to just have HCD literature, accessible methods, and case studies. To conclude the book, note that substantial progress in mining HCD may only be reached through integrating it with wider initiatives (e.g., PtD) and by tackling most of the human-centered design barriers presented in Chapter 3.

The structured approach to mining HCD presented in this book will aid the "prevention through design" (PtD) national initiative being undertaken in the USA and similar safe design strategies in Australia, Europe, and elsewhere. For example, the PtD mission is to prevent occupational injuries, illnesses, and fatalities through the inclusion of prevention considerations in all designs that impact workers (Howard, 2008; NIOSH, 2014). This can be achieved by a range of strategies that include:

- Eliminating hazards and controlling risks to operators to an acceptable level. Ideally, this would be undertaken at their source by "designing out" hazards or at least managing risks as early as possible in the life cycle of equipment, products, and workplaces.
- Explicitly including operator safety and health in the design, redesign, and retrofit of new and existing work premises, structures, tools, facilities, equipment, machinery, products, substances, processes, and organization.
- Enhancing the work environment through inclusion of prevention methods in all designs that impact operators and other stakeholders in the work environment.

This PtD initiative has much in common with the mining human-centered design approach presented here, and is highly compatible with the aims of the HCD work in mining. Although the PtD initiative has been in place for several years in the USA and similar safety by design strategies are in place in leading countries globally, the minerals industry is not yet doing everything possible to make sure this approach is being followed. There is still not enough support to inform the mining industry of the benefits or how-to's of designing this way—that is, a concentration on end-user needs, tasks, and work contexts.

It is apparent in many of the investigative reports of accidents and injuries that both the HCD and PtD approaches are neglected (e.g., Burgess-Limerick, 2011). It is important to provide the mining industry with a structured approach and tools to help them get there, and also to help the industry with changing the prevailing engineering-focused mindset by overcoming many human-centered design barriers.

It should be acknowledged that the industry's neglect of HCD is not purposeful; rather, it is largely a product of the characteristics of mining equipment manufacturers, regulatory agencies, and mining companies: this includes a lack of in-house capacity to address human-centered design issues, a lack of knowledge of information existence and how to apply HCD to their designs, and tight deadlines for mandatory design to meet mining disaster mandates and similar.

Despite this, and despite our attempt to address some of the mining HCD barriers introduced in Chapter 3, we submit that the following actions would further support the growth of human-centered design in the minerals industry:

- *Harness the strength of other initiatives, e.g., the PtD strategy in the USA.* As mentioned earlier, the move towards safety by design is beginning to have greater impact in many high hazard work domains, mining included. This has a natural affinity with HCD in terms of the shared objective of designing out hazards by means of explicitly including operator safety and health in design, redesign, and retrofit of equipment and other work systems, processes, and tasks.

- *Encourage mining companies to provide better site access for designers.* Helping designers to actively see how work is routinely conducted will allow a better understanding of the operators, their tasks, and the work context. Often such access is undertaken on an ad hoc basis, whereas a more structured process could be of assistance to all sides. This issue can be particularly acute for smaller and new technology manufacturers who might not have the necessary mine site or corporate contacts.
- *Work with regulators and other government to minimize the impact of legislative barriers.* For example, in the USA, the MSHA approval process for new technologies can sometimes be a barrier that can slow or prohibit the introduction of useful, human centered technologies in mining. Therefore, working together with MSHA can help create a better implementation process.
- *Legislate the requirement for HCD summary reports for use during mining equipment development, implementation, and procurement.* This would mirror human-centered design work for medical devices in the USA, where the FDA require a summary report.
- *Even if HCD summary reports are not possible, continue to better explain the benefits and successes of mining human-centered design.* This is important so that, during procurement, mining customers ask for evidence from manufacturers of HCD being used during design and development of the equipment/technology. A good example of such human element issues to be considered in design is the variability in shapes, sizes, and abilities of operational and maintenance personnel, and how human-centered design can help successfully design for diverse populations: often very different from the designer's own.
- *Confront the conservative nature of mining by developing better mechanisms for mine sites to celebrate the successes of using "unproven" technology.* The case study of CMOC Northparkes is a good example here. Further ways to promote such successes by means of conferences, awards, and industry articles in strongly recommended. Developing cost benefit analyses—so that adding human-centered design is not seen as an overall cost can also be of great assistance here
- *Similarly, build on the opportunities that new mining technologies and automation can offer.* This can mean that mining no longer just proceeds with operations as usual, but instead embraces automation designed from a HCD perspective. This also can include training to cover skill gaps for operating and maintaining new mining technologies/automation. In this regard, HCD can therefore been seen as consistent with the mining "technology push".

- *Where possible, introduce human-centered design and ergonomics into the curriculum of degree courses in design, mining engineering, and similar.* The possible topics on offer in such programs are often already quite diverse; however, the importance of HCD in the development of fit for purpose equipment and technologies needs to be emphasized to academic course leaders. This will help minimize the chance that HCD is seen as unnecessary by designers, and prevent them from relying on their intuition rather than good ergonomics in the design process. This book, ISO 9241-210 (2010), and Horberry et al. (2011) are all good reference sources for such courses. Indeed, the Horberry et al. (2011) book has recently been used as the textbook in two postgraduate human factors in the minerals industry courses offered by the University of Queensland, Australia.
- *Likewise, run short courses and mentoring programs to assist with the lack of trained human-centered design or human factors professionals in mining companies and OEMs.* The professional societies, such as the Human Factors and Ergonomics Society of Australia, can often offer assistance here.
- *Embolden larger mining OEMs to better communicate HCD both within their businesses and with customers.* One avenue for this is to better establish "voice of the customer" and other equipment feedback processes.
- *Rather than trying to introduce HCD across the whole worldwide mining industry, instead focus on key priority areas for initial human-centered design initiatives.* Such priority areas might be determined by risk—for example, access and egress from surface mobile mining equipment.
- *Further persuade funding bodies to better support the global drive for HCD Research and Development in mining.* The case studies and other successful HCD examples presented in this book will be able to provide good ammunition to show the overall worth of human-centered design in mining.

With these actions, it is anticipated that safe, effective, and fit-for-purpose mining equipment that is suited to both operator needs and operational demands can become a reality.

References

American National Standards Institute [ANSI]. (2011). *Prevention through Design: Guidelines for Addressing Occupational Hazards and Risks in Design and Redesign Processes.* ANSI/ASSE Z590.3-2011.

Australian Standard 4024.1704. (2006). *Safety of Machinery.* http://infostore.saiglobal.com/store/details.aspx?ProductID=308155.

Barone, T.L., Patts, J.R., Janisko, S.J., Colinet, J.F., Patts, L.D., Beck, T.W., and Mischler, S.E. (2015). Sampling and analysis method for measuring airborne coal dust mass in mixtures with limestone (rock) dust, *Journal of Occupational and Environmental Hygiene*, DOI:10.1080/15459624.2015.1116694.

Bernard, B.P. (Ed.) (1997). *Musculoskeletal Disorders and Workplace Factors: A Critical Review of Epidemiologic Evidence for Work-related Disorders of the Neck, Upper Extremities, and Low Back.* US Department of Health and Human Services, National Institute of Occupational Safety and Health. DHHS (NIOSH) Publication No. 97-141.

Booher, H. (2003). *Handbook of Human Systems Integration.* John Wiley & Sons, Hoboken, NJ.

Bovenzi, M. and Hulshof, C.T.J. (1998). An updated review of epidemiologic studies on the relationship between exposure to whole-body vibration and low back pain. *Journal of Sound and Vibration*, 215, 595–611.

Brown, O. (1993). On the relationship between participatory ergonomics, performance and productivity in organisational systems. In: Marras, W., Karwowski, W., Smith, J., and Pacholski, L. (Eds) *The Ergonomics of Manual Work.* Taylor & Francis, London.

Brown, O.J. (2005). Participatory ergonomics. In: Stanton, N., Hedge, A., Brookhuis, K., Salas, E., and Hendrick, H. (Eds) *Handbook of Human Factors and Ergonomics Methods.* CRC Press LLC, Boca Raton, FL.

Burgess-Limerick, R. (2005). Reducing injury risks associated with underground coal mining equipment. *Ergonomics Australia*, 19(2), 14–20.

Burgess-Limerick, R. and Steiner, L. (2006). Injuries associated with continuous miners, shuttle cars, load-haul-dump, and personnel transport in New South Wales underground coal mines. *Mining Technology (TIMM A)*, 115, 160–168.

Burgess-Limerick, R. and Steiner, L. (2007). Opportunities for preventing equipment related injuries in underground coal mines in the USA. *Mining Engineering*, 59, 20.

Burgess-Limerick, R., Straker, L., Pollock, C., Dennis, G., Leveritt, S., and Johnson, S. (2007). Participative ergonomics for manual tasks in coal mining. *International Journal of Industrial Ergonomics*, 37, 145–155.

Burgess-Limerick, R., Krupenia, V., Zupanc, C., Wallis, G., and Steiner, L. (2010a). Reducing control selection errors associated with underground bolting equipment. *Applied Ergonomics*, 41, 549–555.

Burgess-Limerick, R., Krupenia, V., Wallis, G., Pratim-Bannerjee, A., and Steiner, L. (2010b). Directional control-response relationships for mining equipment. *Ergonomics*, 53, 748–757.

Burgess-Limerick, R. (2011). Injuries associated with underground coal mining equipment. *The Ergonomics Open Journal*, 4, (Suppl. 2-M1), 62–73.

Burgess-Limerick, R., Cotea, C., Pietrzak, E., and Fleming, P. (2011). Human systems integration in defence and civilian industries. *Australian Defence Force Journal*, 186, 51–60.

Burgess-Limerick, R. (2012). How on earth moving equipment can ISO 2631 be used to evaluate WBV exposure? *Journal of Health and Safety Research and Practice*, 4(2), 13–21.

Burgess-Limerick, R., Joy, J., Cooke, T., and Horberry, T. (2012). EDEEP—An innovative process for improving the safety of mining equipment. *Minerals*, 2(4), 272–282.

Burgess-Limerick, R., Horberry, T., and Steiner, L. (2014). Bow-tie analysis of a fatal underground coal mine collision. *Ergonomics Australia*, 10, 2.

Burgess-Limerick, R. and Lynas, D. (2015). An iOS application for evaluating whole-body vibration within a workplace risk management process. *Journal of Occupational and Environmental Hygiene*, 12, D137–D142.

Burgess-Limerick, R. (2017). Interface design for haul-truck proximity advisory systems. Australian Coal Association Research Program (Project C24028). www.acarp.com.au.

Camargo, H.E., Peterson, J.S., Kovalchik, P.G., and Alcorn, L.A. (2010). Acoustic assessment of pneumatic and electric jackleg drills used in the mining industry. In C.B. Burroughs and G. Maling (Eds) *Proceedings of the National Conference on Noise Control Engineering and 159th Meeting of the Acoustical Society of America*, Baltimore, MA, April 19–21, pp. 1–11.

Cantley, L.F., Taiwo, O.A., Galusha, D., Barbour, R., Slade, M.D., Tessier-Sherman, B., and Cullen, M.R. (2014). Effect of systematic ergonomic hazard identification and control implementation on musculoskeletal disorder and injury risk. *Scandinavian Journal of Work, Environment and Health*, 40(1), 57.

Claasen, C. (2011). Goaf inertisation and sealing utilising methane from in-seam gas drainage system. In *11th Underground Coal Operators' Conference*, University of Wollongong, Wollongong, NSW, pp. 369–374.

Cloete, S. and Horberry, T. (2014). Collision avoidance and semi-automation in electric rope shovel operation. *Ergonomics Australia*, 4, 2.

Cole, D., Rivilis, I., Van Eerd, D., Cullen, K., Irvin, E., and Kramer, D. (2005). Effectiveness of participatory ergonomic interventions: A systematic review. Institute for Work and Health, Toronto.

Cooke, T. (2015). Human factors methods to develop safer mining equipment. Unpublished PhD thesis, the University of Queensland, Australia.

Dennis, G., Burgess-Limerick, R., and Firth, I. (2015). Successfully implementing a global participative ergonomics program across Rio Tinto. In *Proceedings of the International Ergonomics Association Congress*, Melbourne.

Dixon, S., Theberge, N., and Cole, D. (2009). The case of participatory ergonomics. *Relations Industrielles*, 64(1), 50–74.

Durso, F.T., Boehm-Davis, D.A., and Lee, J.D. (2015). A view of human-systems integration from the academy. In D.A. Boehm-Davis, F.T. Durso, and J.D. Lee (Eds) *APA Handbook of Human-Systems Integration* (pp. 5–19). American Psychological Association, Washington, DC.

EDC (Engineering Design Centre, University of Cambridge, UK). (2017). Inclusive design toolkit—Concept design process: Overview. Downloaded on May 31, 2017 from www.inclusivedesigntoolkit.com/GS_overview/overview.html.

Eger, T.R., Salmoni, A.W., and Whissell, R. (2004). Factors influencing load-haul-dump operator line of sight in underground mining. *Applied Ergonomics*, 35, 93–103.

Eger, T., Salmoni, A., Cann, A., and Jack, R. (2006). Whole-body vibration exposure experienced by mining equipment operators. *Occupational Ergonomics*, 6, 121–127.

Eger, T.E., Godwin, A.A., and Grenier, S.G. (2010). Using visibility tools in Classic JACK to assess line-of-sight issues associated with the operation of mobile equipment. *International Journal of Human Factors Simulation and Modelling*, 1(4), 406–419.

Eger, T., Contratto, M., and Dickey, J. (2011). Influence of driving speed, terrain, seat performance and ride control on predicted health risk based on ISO 2631-1 and EU Directive 2002/44/EC. *Journal of Low Frequency Noise Vibration and Active Control*, 30, 291–312.

EMESRT Design Evaluation for EME Procurement (EDEEP) (no date). Accessed December 12, 2017 from www.emesrt.org/emesrt-design-evaluation-for-eme-procurement.

ErgoTMC. (2016). *User Centered Design*. http://ergotmc.gtri.gatech.edu/ddt/User-Centered_ Design/UCD_TestEvaluation_TE_T.htm.

Flach, J.M., Vicente, K.J., Tanabe, F., Monta, K., and Rasmussen, J. (1998). An ecological approach to interface design. In *42nd Annual Meeting of the Human Factors and Ergonomics Society*, HFES, Santa Monica, CA, pp. 295–299.

Food and Drug Administration. (2016a). Applying human factors and usability engineering to medical devices: Guidance for Industry and Food and Drug Administration Staff. Accessed June 2, 2017 from www.fda.gov/downloads/MedicalDevices/.../ UCM259760.pd.

Food and Drug Administration. (2016b). The code of federal regulations (Title 21, Volume 8, revised April 1, 2016). Accessed June 2, 2017 from www.accessdata.fda.gov/scripts/ cdrh/cfdocs/cfcfr/CFRSearch.cfm?fr=820.30.

Giacomin, J. (2012). *What is Human Centred Design?* P & D Design 2012 (10° Congresso Brasileiro de Pesquisa e Desenvolvimento em Design), São Luís, Brazil.

Godwin, A.A., Eger, T.R., Salmoni, A.W., and Dunn, P.G. (2008). Virtual design modifications yield line-of-sight improvements for LHD operators. *International Journal of Industrial Ergonomics*, 38, 202–210.

Godwin, A.A. and Eger, T.E. (2012). Evaluating underground mining equipment line-of-sight and augmented vision. EMESRT design evaluation case study 12-02. http:// emerst.org./.

Gray, I. and Wood, J. (2015). Outburst risk determination and associated factors. ACARP project C23014 final report.

Gulliksen, J. and Göransson, B. (2001). Reengineering the systems development process for user centered design. In M. Hirose (Ed.) *Proceedings of INTERACT 2001*, IOS Press, Amsterdam.

Gulliksen, J., Göransson, B., Boivie, I., Blomkvist, S., Persson, J. and Cajander, A. (2003). Key principles for user-centred systems design. *Behaviour and Information Technology*, 22(6), 397–409.

Haims, M. and Carayon, P. (1998). Theory and practice for the implementation of "in-house", continuous improvement participatory ergonomics programs. *Applied Ergonomics*, 29, 461–472.

Haines, H.M. and Wilson, J.R. (1998). *Development of a Framework for Participatory Ergonomics*. Health and Safety Executive, HSE Books, Sudbury, Suffolk.

Hass, E.J. and Rost, K.A. (2015). Integrating technology: Learning from mine worker perceptions of proximity detection systems. Society of Mining Engineers Annual Meeting, Feb 15–18, 2015. Denver, CO.

Hayward, J. (1998). Dartbrook mine—A case study. In N. Aziz (Ed.) *Coal 1998: Coal Operators' Conference*, University of Wollongong and the Australasian Institute of Mining and Metallurgy, Wollongong, NSW, pp. 224–238.

Hermawati, S. and Lawson, G. (2013). User-centred design of virtual training for automotive industries. In M. Anderson (Ed.) *Proceedings of the Contemporary Ergonomics and Human Factors 2013*. Taylor & Francis, Cambridge, UK.

Horberry, T., Larsson, T., Johnston, I., and Lambert, J. (2004). Forklift safety, traffic engineering and intelligent transport systems: A case study. *Applied Ergonomics*, 35(6), 575–581.

Horberry, T., Gunatilaka, A., and Regan, M. (2006). Intelligent transport systems for industrial mobile equipment safety. *Journal of Occupational Health and Safety: Australia and New Zealand*, 22(4), 323–334.

Horberry, T., Burgess-Limerick, R., and Steiner, L. (2011). *Human Factors for the Design, Operation and Maintenance of Mining Equipment*. CRC Press, Boca Raton, FL.

Horberry, T. (2012). The health and safety benefits of new technologies in mining: A review and strategy for designing and deploying effective user-centred systems. *Minerals*, 2(4), 417–425.

Horberry, T. and Cooke, T. (2012). Safe and inclusive design of equipment used in the minerals industry. In P. Langdon, J. Clarkson, P. Robinson, J. Lazar, and A. Heylighen (Eds) *Designing Inclusive Systems: Designing Inclusion for Real-World Applications*, Springer-Verlag, London, UK.

Horberry, T. and Lynas, D. (2012). Human interaction with automated mining equipment: The development of an emerging technologies database. *Ergonomics Australia*, 8(1), 1–6.

Horberry, T. and Cooke, T. (2014). Operator acceptance of new technology for industrial mobile equipment. In M. Regan et al. (Ed.) *Driver Acceptance of New Technology*. Ashgate, Oxon, UK.

Horberry, T., Burgess-Limerick, R., Storey, N., Thomas, M., Ruschena, L., Cook, M., and Pettitt, C. (2014). A user-centred safe design approach to control. In the *SIA OHS Body of Knowledge*. www.ohsbok.org.au/wp-content/uploads/2013/12/34.1-User-centredsafe-design-approach-to-control.pdf.

Horberry, T. and Burgess-Limerick, R. (2015). Applying a human-centred process to re-design equipment and work environments. *Safety*, 1, 7–15.

Horberry, T., Burgess-Limerick, R., Cooke, T., and Steiner, L. (2016). Improving mining equipment safety through the use of human centered design. *Ergonomics in Design*. DOI:10.1177/1064804616636299.

Horberry, T., Young, K., and Burgess-Limerick, R. (2016). Proximity warning system interfaces for mining vehicles: Can the minerals industry learn from the automotive domain? In *Proceedings of the International Conference on Ergonomics and Human Factors* (pp. 221–227). Chartered Institute of Ergonomics and Human Factors, Daventry, UK.

Horswill, M.S., Hill, A., and Wetton, M. (2015). Can a video-based hazard perception test used for driver licensing predict crash involvement? *Accident Analysis and Prevention*, 82, 213–219.

Howard, J. (2008). Prevention through design—Introduction. *Journal of Safety Research*, 39, page 113.

Human Centered Strategies. (2006). Human centered design. Downloaded on May 15, 2014 from www.humancenteredstrategies.com/process.php.

International Standards Organisation. (1997). *Evaluation of Human Exposure to Whole-body Vibration. Part 1-General Requirements*. ISO 2631-1. ISO, Geneva, Switzerland.

International Standards Organisation. (2010). *Evaluation of Human Exposure to Whole-body Vibration. Part 1-General Requirements*. ISO 2631-1. Amendment 1. ISO, Geneva, Switzerland.

ISO 9241. (2010). *Ergonomics of Human-System Interaction—Part 210: Human-Centred Design for Interactive Systems*. www.iso.org/iso/catalogue_detail.htm?csnumber=52075.

ISO 12100. (2010). *Safety of Machinery—General Principles for Design—Risk Assessment and Risk Reduction*. www.iso.org/iso/catalogue_detail.htm?csnumber=51528.

ISO 6682. (2008). *Earth-Moving Machinery—Zones of Comfort and Reach for Controls*. www.iso.org/iso/catalogue.

Keith, S.E. and Brammer, A.J. (1994). Rock drill handle vibration: Measurement and hazard estimation. *Journal of Sound and Vibration*, 174, 475–491.

Kirlik, A. (1993). Modelling strategic behavior in human-automation interaction: Why an "aid" can (and should) go unused. *Human Factors*, 35, 221–242.

Kirwan, B. and Ainsworth, L.K. (1992). *A Guide to Task Analysis*. Taylor & Francis, London, UK.

Kowalski-Trakofler, K.M. and Vaught, C. (2012). Psycho-social issues in mine emergencies: The impact on the individual, the organisation and the community. *Minerals*, 2, 129–168.

Kujala, S. (2003). User involvement: A review of the benefits and challenges. *Behaviour and Information Technology*, 22(1), 1–16. DOI:10.1080/01449290301782.

Lee, J.D. and Seppelt, B.D. (2009). Human factors in automation design. In S. Nof (Ed.) *Springer Handbook of Automation*. Springer, New York.

Liker, J.K., Nagamachi, M., and Lifshitz, Y.R. (1989). A comparative analysis of participatory ergonomics programs in U.S. and Japan manufacturing plants. *International Journal of Industrial Ergonomics*, 3, 185–199.

Lynas, D. and Horberry, T. (2010). Exploring the human factors challenges of automated mining equipment. In *Proceedings of the 46th Annual Conference of the Human Factors and Ergonomics Society of Australia (HFESA 2010)*, Sunshine Coast, QLD.

Lynas, D. and Horberry, T. (2011a). Human factors issues with automated mining equipment. *Ergonomics Open*, 4, 74–80. DOI:10.2174/1875934301104010074.

Lynas, D. and Horberry, T. (2011b). A review of Australian human factors research and stakeholder opinions regarding mines of the future. *Ergonomics Australia—HFESA 2011*, 11, 13.

McPhee, B., Foster, G., and Long, A. (2009). *Bad Vibrations*, 2nd ed. Coal Services Health & Safety Trust, Sydney, NSW.

MDG 15. (2002). *Guideline for Mobile and Transportable Equipment for Use in Mines*. Produced by Mine Safety Operations, New South Wales Department of Primary Industries (Australia). www.resourcesandenergy.nsw.gov.au/__data/assets/pdf.../MDG-15.pdf.

Mine Improvement and New Emergency Response Act (MINER Act). (2006). United States Department of Labor. www.msha.gov/MinerAct/MinerActSingleSource.asp.

National Highway Traffic Safety Administration (NHTSA). (2013). U.S. Department of transportation releases policy on automated vehicle development. Downloaded on September 1, 2016 from www.nhtsa.gov/About+NHTSA/Press+Releases/U.S.+Department+of+Transportation+Releases+Policy+on+Automated+Vehicle+Development.

National Research Council. (2013). *Improving Self-Escape for Underground Coal Mines*. National Academies Press, Washington, DC. www.nap.edu/catalog/18300/improving-self-escape-from-underground-coal-mines.

NIOSH. (1974). NIOSH criteria for a recommended standard: Occupational exposure to crystalline silica. U.S. Department of Health, Education, and Welfare, Public Health Service, Center for Disease Control, National Institute for Occupational Safety and Health, DHEW (NIOSH) Publication No. 75–120, Cincinnati, OH.

NIOSH. (2002). NIOSH hazard review: Health effects of occupational exposure to respirable crystalline silica. In P.A. Schulte, F.L. Rice, R.J. Key-Schwartz, D.L. Bartley, P. Baron, P.C. Schlecht, M. Gressel, and A.S. Echt (Eds). U.S. Department of Health and Human Services, Centers for Disease Control and Prevention, National Institute for Occupational Safety and Health, DHHS (NIOSH) Publication No. 2002–129, Cincinnati, OH.

NIOSH. (2012). Dust control handbook for industrial minerals mining and processing. In A.B. Cecala, A.D. O'Brien, J. Schall, J.F. Colinet, W.R. Fox, R.J. Franta, J. Joy, W.R. Reed, P.W. Resser, J.R. Rounds, and M.J. Schultz (Eds). U.S. Department of Health and Human Services, Centers for Disease Control and Prevention, National Institute for Occupational Safety and Health, DHHS (NIOSH) Publication No. 2012–112 (RI 9689), 1–284, Pittsburgh, PA.

NIOSH, CDC. (2014). The State of the National initiative on prevention through design. DHHS (NIOSH) Publication No. 2014–123.

O'Sullivan, J. (2007). Ergonomics in the design process. *Ergonomics Australia*, 21(2), 13–20.

Onal, E., Craddock, C., Endsley, M.R., and Chapman, A. (2013). From theory to practice: Designing for situation awareness to transform shovel operator interfaces, reduce costs, and increase safety. *Canadian Institute of Mining, Metallurgy and Petroleum (CIM) Journal*, 4(4). http://store.cim.org/en/from-theory-to-practice-designing-for-situation-awareness-to-transform-shovel-operator-interfaces-reduce-costs-and-increase-safety.

Parasuraman, R. and Riley, V. (1997). Humans and automation: Use, misuse, disuse, abuse. *Human Factors*, 39, 230–253.

Pazell, S., Burgess-Limerick, R., Horberry, T, Dennis, G., and Wakeling, C. (2016). RIO TINTO WEIPA: The value proposition of good work design. Health, safe and productive by design. In *Proceedings of the 51st Annual Conference of the Human Factors and Ergonomics Society of Australia*, Gold Coast, QLD, November 6–9, 2016.

Pike River. (2012). Royal Commission on the Pike River coal mine tragedy. Final report. Wellington, New Zealand. ISBN 978-0-477-10378-7.

Price, D.L. (1991). Demonstration of continuous cutting/bolting/rib-bolting development machine. [The Voest-Alpine ABM 20 Bolter Miner]. Australian Coal Association Research Program report 1032.

Rivilis, I., Van Eerd, D., Cullen, K., Cole, D.C., Irvin, E., Tyson, J., and Mahood, Q. (2008). Effectiveness of participatory ergonomic interventions on health outcomes: A systematic review. *Applied Ergonomics*, 39(3), 342–358.

Rouse, W. (2007). *People and Organizations: Explorations of Human-Centered Design*. John Wiley & Sons, Hoboken, NJ.

Sammarco, J., NIOSH PMRD, CDC (no date). *Adequate Underground Lighting*. www.cdc.gov/niosh/mining/topics/Illumination.html.

Sammarco, J., Reyes, M.A., and Gallagher, S. (2009). Do light-emitting diode cap lamps enable improvements in miner safety? *Mining Engineering*, 61(10), 43.

Sandvik (no date). Accessed December 12, 2017 from http://sandvikmakeitcount.com/safety/sandvik-edeep-process and http://sandvikmakeitcount.com/safety/sandvik-edeep-process/dd422i-finland.

Schulte, P., Rinehart, R., Okun, A., Geraci, C., and Heidel, D. (2008). National prevention through design (PtD) initiative. *Journal of Safety Research*, 39, 115–121.

Sharples, S., Worthy, T., and Brown, F. (2016). *The Human Connection: How Ergonomics and Human Factors Can Improve Lives, Business and Society*. Chartered Institute of Ergonomics and Human Factors, Loughborough, UK.

Silverstein, B. and Clark, R. (2004). Interventions to reduce work-related musculoskeletal disorders. *Journal of Electromyography and Kinesiology*, 14, 135–152.

Simpson, G., Horberry, T., and Joy, J. (2009). *Understanding Human Error in Mine Safety*. Ashgate Press, London, UK.

Society of Automotive Engineers. (2014). *Taxonomy and Definitions for Terms Related to on-Road Motor Vehicle Automated Driving Systems*. Standard number J3016_201401. www.standards.sae.org.

Stanton, N., Salmon, P., Rafferty, L.A, Walker, G., Baber, C., and Jenkins, D. (2013). *Human Factors Methods: A Practical Guide for Engineering and Design*, 2nd edition. Ashgate Press, London, UK.

Steiner, L. (2014). Reducing underground coal roof bolting injury risks through equipment design. Unpublished PhD thesis, the University of Queensland, Australia.

Steiner, L. and Burgess-Limerick, R. (2013). Shape-coding and length-coding as a measure to reduce the probability of selection errors during the control of industrial equipment. *IIE Transactions on Occupational Ergonomics and Human Factors*, 1, 224–234.

Steiner, L., Burgess-Limerick, R., Eiter, B., Porter, W., and Matty, T. (2013). Visual warning system to reduce errors while operating roof bolting machines. *Journal of Safety Research*, 44, 37–44.

Steiner, L., Burgess-Limerick, R., and Porter, W. (2014). Directional control-response compatibility relationships assessed by physical simulation of an underground bolting machine. *Human Factors*, 56, 384–391.

Straker, L., Burgess-Limerick, R., Egeskov, R., and Pollock, C. (2004). A randomised and controlled trial of a participative ergonomics program (PErforM). *Ergonomics*, 47, 166–188.

Tichon, J. (2011). Training. In T. Horberry, R. Burgess-Limerick, and L. Steiner (Eds) *Human Factors for the Design, Operation and Maintenance of Mining Equipment*. CRC Press, Boca Raton, FL.

Torma-Krajewski, J., Hipes, C., Steiner, L., and Burgess-Limerick, R. (2007). Ergonomic interventions at Vulcan Materials Company. *Mining Engineering*, 59(11), 54–58.

Travis, D. (2009). *User-Centered Design: The Fable of the User-Centered Designer.* www.userfocus.co.uk/fable/.

US Department of Defense (DOD). (2008). Operation of the defense acquisition system. Instruction number 5000.02.

Van der Laan, J.D., Heino, A., and De Waard, D. (1997). A simple procedure for the assessment of acceptance of advanced transport telematics. *Transportation Research Part C: Emerging Technologies*, 5(1), 1–10.

Vink, P., Peeters, M., Grundemann, R.M.W., Smulders, P.G.W., Kompier, M.A.J., and Dul, J. (1995). A participatory ergonomics approach to reduce mental and physical workload. *International Journal of Industrial Ergonomics*, 15, 389–396.

Weeks, J.L. and Rose, C. (2006). Metal and non-metal miners' exposure to crystalline silica, 1998–2002. *American Journal of Industrial Medicine*, 49, 523–534.

Wester, J. and Burgess-Limerick, R. (2015). Using a task-based risk assessment process (EDEEP) to improve equipment design safety: A case study of an exploration drill rig. *Mining Technology*, special issue, 124, 69–72.

Wiener, E.L. (1989). Human factors of advanced technology ("glass cockpit") transport aircraft. (NASA Contractor Rep. 177528). NASA-Ames Research Center.

Williams, J.C. (1986). HEART A proposed method for assessing and reducing human error. In *9th Advanced in Reliability Technology Symposium*, University of Bradford, Bradford, UK.

Wilson, J.R. (1995). Ergonomics and participation. In J.R. Wilson and E.N. Corlett (Eds) *Evaluation of Human Work*, 2nd ed. Taylor & Francis, London.

Wolfgang, R. and Burgess-Limerick, R. (2014a). Whole-body vibration exposure of haul truck drivers at a surface coal mine. *Applied Ergonomics*, 45, 1700–1704.

Wolfgang, R. and Burgess-Limerick, R. (2014b). Using consumer electronic devices to estimate whole-body vibration exposure. *Journal of Occupational and Environmental Hygiene*, 11(6), D77–D81.

Wolfgang, R., Di Corletto, L., and Burgess-Limerick, R. (2014). Can an iPod Touch be used to assess whole-body vibration associated with mining equipment? *The Annals of Occupational Hygiene*, 58, 1200–1204.

Woods, D. and Hollnagel, E. (2005). *Joint Cognitive Systems.* CRC Press, Boca Raton, FL.

Xiao, T., Horberry, T., and Cliff, D. (2015). Analysing mine emergency management needs: A cognitive work analysis approach. *International Journal of Emergency Management*, 11(3), 191–208.

Yenchek, M.R. and Sammarco, J.J. (2010). The potential impact of light emitting diode lighting on reducing mining injuries during operation and maintenance of lighting systems. *Safety Science*, 48(10), 1380–1386.

Index

Page numbers followed by *f* indicate figures; those followed by *t* indicate tables.